Sea Level Rise

This is Volume 75 in the
INTERNATIONAL GEOPHYSICS SERIES
A series of monographs and textbooks
Edited by RENATA DMOWSKA, JAMES R. HOLTON, and
H. THOMAS ROSSBY

A complete list of books in this series appears at the end of this volume.

Sea Level Rise
History and Consequences

Bruce C. Douglas*,†
Michael S. Kearney*
Stephen P. Leatherman†

*DEPARTMENT OF GEOGRAPHY
UNIVERSITY OF MARYLAND
COLLEGE PARK, MARYLAND

†INTERNATIONAL HURRICANE CENTER
FLORIDA INTERNATIONAL UNIVERSITY
MIAMI, FLORIDA

ACADEMIC PRESS

A Harcourt Science and Technology Company

San Diego   San Francisco   New York
Boston   London   Sydney   Tokyo

Copyright © 2001 by ACADEMIC PRESS

Academic Press
A Harcourt Science and Technology Company
525 B Street, Suite 1900, San Diego, California 92101-4495, USA
http://www.academicpress.com

Academic Press
Harcourt Place, 32 Jamestown Road, London NW1 7BY, UK
http://www.academicpress.com

Library of Congress Catalog Card Number: 00-104280

International Standard Book Number: 0-12-221345-9

PRINTED IN THE UNITED STATES OF AMERICA
00 01 02 03 04 05   MM   9 8 7 6 5 4 3 2 1

Sharp's Island, Maryland, ca. 1950. This photo shows what remained of an island that probably was approximately 700 acres in size at the time of original settlement in the late 17th century, and that still covered almost 600 acres in 1850. Until the first decades of the 20th century, Sharp's Island supported several large farms (at least one of 300 acres) and a hotel until 1910. Today, the island has disappeared, with only the historic Sharp's Island Light marking its former position. (Photo used with permission from Douglas Hanks, Jr.)

Contents

Chapter 1 An Introduction to Sea Level

Bruce C. Douglas

Chapter 2 Late Holocene Sea Level Variations

Michael S. Kearney

Chapter 6 Observations of Sea Level Change from
Satellite Altimetry

R. Steven Nerem and Gary T. Mitchum

Chapter 7 Decadal Variability of Sea Level

W. Sturges and B. G. Hong

Chapter 8 Social and Economic Costs of Sea Level Rise

Stephen P. Leatherman

Contributors

Numbers in parentheses indicate page numbers on which authors' contributions begin.

Bruce C. Douglas (1, 37) Department of Geography, University of Maryland, College Park, Maryland 20742; and Laboratory for Coastal Research, Florida International University, Miami, Florida 33199

Vivien Gornitz (97) Center for Climate Systems Research, Columbia University, New York, New York 10027; and Goddard Institute for Space Studies, NASA, New York, New York 10025

B. G. Hong (165) Department of Marine Science, University of South Florida, St. Petersburg, Florida 33716

Michael S. Kearney (13) Department of Geography, University of Maryland, College Park, Maryland 20742

Stephen P. Leatherman (181) Laboratory for Coastal Research & International Hurricane Center, Florida International University, Miami, Florida 33199

Gary T. Mitchum (121) Department of Marine Science, University of South Florida, St. Petersburg, Florida 33701

R. Steven Nerem (121) Center for Space Research, The University of Texas at Austin, Austin, Texas 78759-5321

W. R. Peltier (65) Department of Physics, University of Toronto, Toronto, Ontario, Canada M5S 1A7

W. Sturges (165) Department of Oceanography, Florida State University, Tallahassee, Florida 32306

Foreword

This book describes both clearly and in detail the complexity behind the deceptively simple subject of sea level rise, a topic of considerable scientific interest and increasing economic importance. The concept of sea level rise is quite straightforward. Some 97% of all the water on Earth is now in the oceans; most of the rest is found in glaciers, much of it in Antarctica and Greenland. Some 20,000 years ago at the peak of the last ice age, much more water was in ice and the sea level was more than 100 meters lower than it is today. The glaciers began to melt, the oceans began to fill, and the shorelines were pushed back as the sea level rose. The process continues, and the results are obvious. Archeologists don Aqua-lungs and explore the ancient port of Alexandria. Closer to home and more recent in time, St. Clements island in the Potomac River was a heavily wooded 160 hectares when first occupied by Virginia colonists. Today, some 350 years later it is about 16 hectares and has little in the way of vegetation. Pictures of battered beach houses and hotels eroded by waves after a particularly vicious winter storm moves up the east coast of the United States are a regular feature of our television news.

In the early 1980s when the issue of global warming first grabbed the headlines, I was in Washington as head of the National Oceanic and Atmospheric Administration. One of NOAA's tasks is to predict the tides and maintain this nation's vast array of tide gauges. What effect, I was asked, will global warming have on the change in sea level? It was embarrassing to admit that we really could not say much more than is in the above paragraph. Yes, sea level has risen in the past; we assume it still is rising, but uncertainty remains about how fast it has been rising recently, and thus we are not in a very good position to estimate how fast it might rise in the future. As this volume attests, there continue to be a number of perplexing issues. There is still uncertainty in some areas, but we do know so much more, not just about the changing volume of the ocean, but about yearly and regional variations in sea level and the reasons for them. Even more exciting, the technology now available suggests that we will soon know very much more.

We now believe that the ocean volume has been increasing since the middle of the 19th century at a rate equivalent to raising sea level almost 2 mm/yr, a rate considerably faster than that for the previous thousand years, although how much faster is subject to some uncertainty. That this increase in the rate of sea level rise began well before the rise in our mean atmospheric temperature of recent years gives pause to those who wish to assign its cause to anthropogenic-driven global warming.

Tracking the changing volume of the waters of the ocean, as distinguished from measuring local sea level, is not a simple task. Many traps lie in wait for the unwary, and not long ago many who had examined the problem were skeptical that we would ever achieve useful quantitative information. It has not been easy. Do not expect to examine a half-dozen years of local tide gauge records and derive a useful value. First, at least a 50-year record is required, because there are year-to-year changes (some of which we understand, but a number of which we do not) which are likely to bias records that are much shorter. Second, and often more difficult to resolve, the land bordering the sea also moves up and down. In much of Scandinavia the local sea level is dropping because the land is rising (several millimeters a year in places), continuing to rebound from the heavy weight of the glaciers removed several thousand years ago. But isostatic adjustment in those areas formerly under the ice requires some form of viscoelastic compensation in those areas away from the former ice sheets. For example, even if there were no change in the volume of the oceans, we now believe that the sea level would be rising along the east coast of the United States at about 1.4 mm/yr because that is the rate the earth is sinking in this part of the world. As a consequence the actual rise of sea level in this region is nearly double that caused by the change in the ocean volume.

Why is the volume of the oceans increasing? Do not expect to find an unambiguous answer in this book. Perhaps it is the melting of the last of our major ice fields. That is certainly what many believe, but we do not have sufficient information about the volume of ice on either Greenland or Antarctica, let alone its rate of change, to give an unambiguous answer. Perhaps the ocean is getting slightly warmer. If it is, then seawater will expand, and the volume of the ocean will increase although its mass will remain unchanged. An increase of the average ocean temperature (top to bottom) of only a few hundredths of a degree per year is all that is required to raise the sea level a couple of millimeters per year, but we do not have the kind of historical ocean temperature records to either prove or disprove such a possibility.

There may be better data on why humankind's activities of the last half century should be driving sea level lower. We have a good record of the number of dams built in the last half century and the amount of water they control. These dams change the historical flow of water from land to rivers and on to the ocean, and one can make educated guesses whether this should either increase or decrease the rate at which water reaches the ocean. Apparently, the largest single effect is the loss of water from behind the dam which leaches out of the bottom and back into groundwater. This water never makes it to farmland, homes, or industry, nor does it evaporate, later to fall as rain.

This water completely bypasses the ocean. A strong case can be made that the rate at which the volume of dammed water is increasing, and thus the rate at which this water is bypassing the usual cycle, is equivalent to a decrease in sea level of possibly many tenths of a millimeter per year.

The rate of change of sea level varies from year to year and place to place. Evidence of past El Niños can clearly be seen in the long-term tidal records of San Diego and San Francisco. Year-to-year changes in the intensity of the wind-driven circulation in the North Atlantic are captured in the yearly changes in mean sea level recorded by tide gauges along the U.S. east coast. With the significant increase in tide gauge accuracy, not the least of which is the removal of the earth movement problem with the availability of GPS, one can expect tide gauges to contribute to an ever-increasing array of geophysical problems.

And finally, if sea level continues to rise, if it is indeed rising at a more rapid rate now than it was a century ago, and if, as some suggest, that rate of rise will increase as a consequence of global warming, what effect will this rising sea level have on society? To those who live in the Ganges delta of Bangladesh, on coral atolls in the Pacific, or below sea level in The Netherlands, this subject holds special interest. One estimate has some 100 million of us living within one meter of sea level. I expect they will be among those most interested in the latest news on this subject.

John A. Knauss

Preface

One of the more serious and certain consequences of global warming is an increase in sea level. Intense concern is expressed for coastal regions of the planet where much of the world's population is found, because of the disastrous flooding that would follow from rapid melting or collapse of a major polar ice sheet. There is by contrast far less public consciousness of the rise in sea level that has occurred during the past 100 years and its very significant environmental and economic impacts. This is unfortunate, because even if sea level in the future rises at only the present rate, very severe costs will ensue because of the rapid economic development and population expansion occurring in the world's coastal regions. This "collision course" of social and sea level trends was largely the motivation for this book.

In the scientific community there is a growing consensus that the overall worldwide rate of sea level rise during the 20th century has been nearly 2 mm per year, sharply (10-fold) higher than the average of the past several millennia. This result is based on analyses of tide gauge data taken since the late 19th century, historical land records, and geological evidence from the late Holocene period. Apparently global sea level has a natural temporal variability that ranges from seasonal to millennial unrelated to any additional variations that might arise from anthropogenic changes of the atmosphere.

This book is designed to be suitable for a senior- or first-year graduate-level course in the earth or environmental sciences, including geography, geology, and marine science. Special emphasis is given to the evidence for historical sea level change, and case studies are used to demonstrate the resulting consequences. The emphasis is placed on the concepts involved in the causes and consequences of sea level change. Nevertheless, the presentation is not "watered down" to the merely descriptive. Enough detail is given so that the concepts can be understood and the many references profitably consulted. We make extensive use of graphical presentations, believing that graphics and data are the best teachers. The precise answers to the geophysical questions may change as a result of new data and research, but not the questions that must be asked if the accelerating increase in coastal population and development is to be managed in a reasonable and environmentally sound manner. We hope that this volume will provide a good basis for further research by its readers, and at the same time provide those interested in policy questions an overview of the subject that will serve them well in their work.

The first chapter of this book is introductory. The concept of sea level and its use as an elevation reference are discussed, and examples of the temporal and geographic variations of contemporary sea level change are presented. Sea level and its variations are surprisingly complicated, being affected by a large number of local, regional, and global factors. The impact of sea level change on the coastal landscape is also introduced, particularly in regard to beach erosion and coastal development.

Chapter 2 covers sea level change during the past 8–10 millennia, but before the era of the recording tide gauge. The sea level record during this time is obtained from geological, geochemical, and marine organism analyses and shows what some investigators believe to be significant variations possibly related to climate fluctuations. The chapter considers in much detail the evidence for intercentennial fluctuations of sea level on the order of a few meters that have occurred during the late Holocene and their possible relation to paleoclimate.

Chapter 3 is concerned with monitoring sea level in the era of the recording tide gauge, which began near the middle of the 19th century. A tide gauge is conceptually just a fancy dipstick, but it records, in addition to the global rate of sea level rise, effects of ocean circulation, meteorological forcing, and local or regional uplift or subsidence at the measurement site. In addition, a tide gauge site must provide a highly accurate and precise data record for scores of years in spite of repairs, replacements, and changes in technology. The international effort to provide long and high-quality sea level records has provided one of the essential indicators of contemporary climate change, including the El Niño phenomenon. But the existence of sea level records is not enough. The problems involved in analyzing and interpreting tide gauge records, and deciding which ones are actually useful for climate analyses, are presented by way of examples of particular water level series. Methodology is an important issue, because researchers have obtained very different results for global sea level rise during the last several decades, in spite of their use of a common data base of sea levels. This chapter also examines the issue of whether an acceleration of sea level has occurred in the last 100 years and provides a list of tide gauge records and trends suitable for determinations of the 20th-century rate of global sea level rise.

The fourth chapter of the book considers in some detail the extended geophysical response of the earth to the melting of the great ice sheets that began at the time of the last glacial maximum about 21,000 calendar years ago. The time after 4000 years ago, when glacial melting was completed, is especially interesting and important for investigations of modern sea level rise. This is so because relative sea levels continued to change in response to removal of the ice load, a process ongoing to this day at rates significant compared to measured trends of sea level. The new results in Chapter 4 yield corrections for glacial isostatic adjustment (GIA) for present-day observed sea level trends, enabling computation of an accurate estimate of 20th century global sea level rise. A convincing case is made that the 20th century rate of rise must be in the range 1.8–1.9 mm/yr, rather than at the much lower values suggested by some other investigators.

Chapter 5 covers a topic that seems implausible at first thought—the effect on the rate of sea level rise due to impoundment of water in large and small reservoirs and groundwater mining. This topic has many facets (some of which have a fair degree of uncertainty), but it appears possible that a significant increase in the rate of global sea level rise during the past 50 years has been masked by anthropogenic alterations to the hydrological cycle.

Chapter 6 presents an exciting new technology: the direct measurement of sea level by an Earth-orbiting satellite. It sounds improbable that a satellite-borne instrument could detect variations of global average sea level to an accuracy on the order of a millimeter per year, but the results obtained thus far are persuasive. Satellite altimetry has provided a breakthrough technology for observing seasonal–interannual El Niño events and may well achieve the same result for global sea level change. It will be an important milestone for climate science to have an estimate of the rate of sea level rise from altimetric satellites for the decades immediately before and after the change of the millennium for comparison to the rate for the entire 20th century.

The seventh chapter is concerned with explaining the prominent seasonal-to-interdecadal variations of sea level. These large fluctuations are highly variable regionally and are large enough to obscure an underlying trend of sea level for at least 50 years or more. Results indicate that it is possible to calculate the amplitude and phase of these variations from meteorological data to a high degree of accuracy for some sites. This creates the possibility of obtaining much "cleaner" records of long-term sea level change and earlier detection of an acceleration of the rate that can be used to confirm satellite altimetric estimates.

Chapter 8 deals with the impact of sea level rise on coastal habitability. Land is lost directly by inundation and over the long term by coastal erosion that takes place at a rate several orders of magnitude greater than the rate of sea level rise. Sea level rise over an extended period also exacerbates the damage caused by severe storms since the energetic storm-driven waves reach farther inland. We present the evidence for sea level rise as the fundamental driving force for long-term beach erosion and wetlands loss, and consider the consequences for highly developed coastal regions and island nations.

Included with this book is a CD-ROM containing monthly mean tide gauge data from the Permanent Service for Mean Sea Level. Corrections for GIA calculated by W. R. Peltier are given for each site to facilitate correcting sea level trends for this important effect. Also included is an animated presentation of global sea surface temperature anomalies during the years 1982–1999 which vividly illustrates the two largest El Niño events of the 20th century. The material on the CD-ROM is in the form of either text files or Web sites that can be opened by widely available Web browsers.

Chapter 1 | An Introduction to Sea Level

Bruce C. Douglas

1.1 THE IMPORTANCE OF SEA LEVEL RISE

Compared to spectacular geophysical phenomena such as earthquakes, volcanoes, and hurricanes, the contemporary rate of long-term sea level rise appears benign. For most populated regions of the world, sea level is increasing at a rate of only a few millimeters per year. There may be an additional component of rise from vertical land movement that locally alters the rate from the global value, but usually by no more than a factor of 2 or 3. Local sea level actually falls at some sites in tectonically active regions of the earth (e.g., in Japan and Alaska) or areas in the Baltic deeply covered by ice sheets at the last glacial maximum 21,000 years ago. However, these are the exceptions. The situation in the conterminous United States is more typical, where the local sea level is rising virtually everywhere at rates averaging about 2–3 mm/yr, with rates up to 6 mm/yr or more in some specific areas.

What makes rising water level so important to humanity is its potential to alter ecosystems and habitability in coastal regions, where an ever-increasing percentage of the population is located. In the United States, a majority of the populace live in the 25 states having coastlines on the Gulf of Mexico and the Atlantic and Pacific Oceans. It is further estimated that about one-half of this population lives within 50 miles of a shoreline (NRC, 1990). Considering the entire planet, Cohen *et al.* (1997) estimated that in 1994 about 2.1 billion people (37% of the world's population) lived within 100 km of a coast. For many persons, such as those living in the Bay of Bengal, sea level rise is a life-and-death issue because as sea level rises, vulnerability to severe storms also increases.

Nicholls and Leatherman (1994) summarize the physical effects of sea level rise in five categories. These are erosion of beaches and bluffs, increased flooding and storm damage, inundation of low-lying areas, salt intrusion into aquifers and surface waters, and higher water tables. The first of these impacts is the most familiar. Beach erosion is ubiquitous in the United States and is also a problem worldwide. Along the U.S. east coast about 80–90% of the beaches are eroding due to long- and short-term processes (Galgano, 1998;

Galgano *et al.,* 1998). Bruun (1962) long ago proposed that long-term erosion of sandy beaches was a consequence of sea level rise. His model predicts that beaches will erode by 50–200 times the rate of increase of sea level, a far greater effect than simple inundation. Verification of the Bruun theory was achieved in a small wave tank by Schwarz (1965, 1967) and for areas in Lake Michigan (Hands, 1983) where the tides are negligible and very large variations of water level occur in relatively short times. But validation of the concept for open ocean coastlines with their complex interplay of geological, meteorological, and oceanographic factors has remained elusive and controversial (Pilkey and Davis, 1987). However, Zhang (1998) and Leatherman *et al.* (2000) have demonstrated a relationship between sea level rise and beach erosion by analyzing a new large database of historical shoreline positions and sea level rise computed from tide gauge measurements. It has been found that open-ocean sandy beaches on the U.S. east coast not affected by inlets or engineering modifications erode at a rate that averages about 150 times the rate of sea level rise. There is no reason to believe that this result is not a general one for sandy beaches everywhere. As an example of the importance of erosion generated by sea level rise, consider the popular middle Atlantic beach resort of Ocean City, Maryland. The rate of sea level rise in the 20th century in this area has been about 3.5 mm/yr, which translates into a beach erosion rate of about 5 m per decade. Since the beach has a nominal width of only a few tens of meters, sand replenishment (often known as beach nourishment) is frequently required to provide a beach wide enough for recreational use and for protection of buildings from storm-generated waves and surges. If the rate of rise of sea level increases as a consequence of global warming, the already-expensive ($82 million to date for this community) beach nourishment program will become an even larger financial burden.

Rising sea levels have other important effects besides erosion. Low-lying coastal plains are vulnerable to inundation and suffer serious consequences of salt intrusion into aquifers. Some islands in the Chesapeake Bay that once supported farming and logging activities have become so contaminated with salt that agricultural activities are now impossible. Other islands in the Bay have disappeared (Kearney and Stevenson, 1991). The roughly 10,000 km of coastline in the Bay is eroding rapidly, creating serious economic, social, and political problems for the region. Rising sea levels also threaten coastal ecosystems when marshes drown because they cannot build upward (vertically accrete) fast enough. Marshes develop ponds as they drown, and ultimately can disappear entirely, as is happening now to the important wildlife refuge at Blackwater, Maryland.

Developing countries and island nations are especially at risk from rising sea levels. As a rule, they lack the resources for mitigation activity or relocation of populations to less threatened areas as sea level increases. The very existence of some island nations is in jeopardy. In terms of sheer numbers of affected persons, the greatest problem lies in populated river deltas. About

100 million people live within one meter of present-day mean sea level (Nicholls and Leatherman, 1995). As habitable land in the affected locations dwindles, neighboring areas could experience significant political stress from environmental refugees. These few examples of the effect of sea level rise give only a glimpse of the threat of increasing sea levels to the environment. A more detailed view is the subject of Chapter 8.

In the earth science community there is an increasing consensus that the overall global rate of sea level rise during the last approximately 100 years has been nearly 2 mm/yr (Warrick *et al.*, 1996), sharply higher than the average rate during the last several millennia. This result is based on analyses of tide gauge data taken since the 19th century, historical land records, archeological data, and geological evidence from the late Holocene period. Possibly half of the apparent increase of global sea level during the past 100 years can be accounted for by thermal expansion of the oceans and melting of small glaciers (Warrick *et al.*, 1996). The remainder remains unaccounted for; the obvious candidates are the Antarctic and Greenland ice sheets, but their ice mass balance is too poorly known for a definitive conclusion (Warrick *et al.*, 1996; Bindschadler, 1998; Wingham *et al.*, 1998). It is even possible that an *additional* rise of nearly 1 mm/yr during the last 50 years has been averted by net above-ground storage of water in large and small reservoirs, as discussed in Chapter 5. In any case, although the present (i.e., the past 100 years) rate of rise and its interpretation are subject to a certain amount of disagreement, it is a fact that sea level is rising in most coastal regions just when rapid coastal development is taking place. If global warming and an increased rate of sea level rise occur in the next century (Warrick *et al.*, 1996), present-day problems attributable to sea level increase will be exacerbated.

1.2 SEA LEVEL AND THE GEOID

The planar-appearing ocean surface stands in immense contrast to that of the land, where the tallest mountain peaks are more than 8 km in elevation above sea level. Beneath the ocean surface, similarly tall mountains (called sea mounts) also exist, and the ocean depth at some great trenches is greater than the height of the tallest mountains. The almost featureless ocean surface thus has an obvious appeal as an elevation reference.

The ocean surface does exhibit some height variability. There are heating, circulation, and meteorological effects on the order of 1 m and in most areas the lunar and solar tides create a semi-daily sea level variation of a meter or so. (For a comprehensive treatise on tides, consult Pugh, 1987.) But computation of a useful mean value of sea level can be achieved over an interval by a simple averaging process. There are also trends in the level of the sea, but as noted, the long-term trend of sea level at the vast majority of the world's coastlines is at present only a few millimeters per year. This rate is so small that for some purposes, such as navigation, it is often ignored.

Using sea level as an elevation reference for the coast has an intuitive appeal, but what is meant exactly by "elevation above sea level" in the middle of a continent? The concept of a level surface gives the answer. A level surface is one on which the potential energy (including the centrifugal potential due to the rotation of the earth) is everywhere the same. There are an infinite number of these surfaces; the one chosen as the reference surface in geodesy is the geoid, which most nearly coincides with the ocean surface. If there were no ocean currents or atmospheric forces on the water, the ocean surface would coincide exactly with the geoid. (If it did not, the water would flow rapidly until it did.) On land the geoid can be thought of as a surface coincident with the water surface on a network of very narrow sea level canals. Then one's elevation above sea level is in essence one's altitude (measured along the line a plumb bob would take) above the surface of a fictitious underfoot sea level canal. The subject of physical geodesy is much concerned with the geoid and determining elevations relative to it from a combination of geometric and gravimetric observations. A curious reader should consult Heiskanen and Moritz (1967), a text highly regarded by geodesists, for the mathematical details of the subject. An example of a computed geoid is shown in Fig. 1.1 (see color plate). Detailed information about it is available at the NOAA Web site www.ngs.noaa.gov. This Web site is a very rich source of information concerning regional and global geoids and their computation and interpretation.

In Fig. 1.1 variations of the geoid are shown with respect to an ellipsoidal reference surface (see below). This map illustrates that the geoid over the United States has variations of a few tens of meters and is correlated with the topography to a great extent. In the conterminous United States, the geoid variations (Fig. 1.1) range from a low of -51.6 m in the Atlantic (magenta) to a high of -7.2 m (red) in the Rocky Mountains. The role played by elevations relative to the geoid is important for a number of scientific and practical concerns. As an example of the latter, if water is to flow in the correct direction in pipes and canals, then the system must have a slope relative to the geoid. Concerning their scientific importance, these geoid variations reflect underlying geophysical phenomena. For an appreciation of the relation between geophysics and geodesy, see the survey volume by Lambeck (1988).

Geoid undulations, such as those shown in Fig. 1.1, can be considered heights of the geoid relative to a "best fitting" reference surface. This reference surface is most commonly an ellipsoid of revolution with the same center of mass and orientation as the geoid. Standard reference surfaces, chosen to most closely approximate the size and shape of the geoid, are selected and sanctioned by international scientific bodies (see Chovitz (1981) for an overview). Over the entire earth, geoid undulations (and hence the mean sea surface) have a root-mean-square variation of about ±35 m, with the most extreme excursions about three times as great. These large undulations are about two orders of magnitude greater than the so-called sea surface topography, the deviations of the sea surface from level (i.e., the geoid) caused by ocean currents and meteorological processes. Satellite altimetry, a relatively new and spectacularly successful measurement technique, has resulted in a revolutionary increase in knowledge of the marine geoid by directly observing the geoidal undulations of the sea surface with a satellite-borne radar ranging system. The technique has also proven capable of measuring the meter-level sea surface topography resulting from ocean currents and their variations and apparently even has the capability to determine the global rate of sea level rise with millimeter accuracy. A discussion of this remarkable technology is reserved for Chapter 6.

As already noted, the sea surface closely approximates the geoid. Deviations from the geoid due to ocean currents and wind forcing are only about 1% ($\approx$ 1 m) of the maximum amplitude of the geoid undulations. Of course, the reason that the sea surface displays so little height deviation from the geoid is that water does not possess any shear strength. The crust of the earth in contrast possesses strength sufficient to support in part or whole the topographic variations of the land. But we shall see later in Chapters 2 and 4 that the earth's crust can deform, and did so by hundreds of meters under the weight of the great ice sheets that existed at the peak of the last glacial maximum 21,000 years ago. Even though those ice sheets melted at least by 5000–4000 BP, the earth did not completely adjust to the removal of the ice load by that time. It is in fact still adjusting from removal of the load by amounts that are significant compared to the contemporary rate of global sea level rise. This means that water level measurements reflect both the global change of sea level and any local subsidence or emergence of the land that is ongoing from the last deglaciation. This continuing glacial isostatic adjustment (GIA) is one of the most important issues in the interpretation of modern sea level records because it is highly variable geographically, and difficult to model. As an illustration, consider Fig. 1.2, which presents a contour plot of 20th century trends of sea level at 70 European sites derived from tide gauge records selected on the basis of data quality. Since there is not coverage over all 360 degrees of azimuth from the load center, contours to the northeast do not close and values there are unrealistic. But the obvious bullseye pattern demonstrates the relict GIA that follows the long-ago completed melting of the Fennoscandian ice sheet. Note that in the Baltic, contemporary sea level is actually *falling* up to 10 mm/yr. This area is the location of the center of the Fennoscandian ice sheet, whose weight caused the earth underneath it to sink and adjacent regions to uplift. Removal of the ice load was completed long ago, but the 20th-century sea level trends show that the area is still rebounding enough to cause a fall in relative sea level. The order of magnitude of isostatic readjustment on modern vertical motion of the land is important

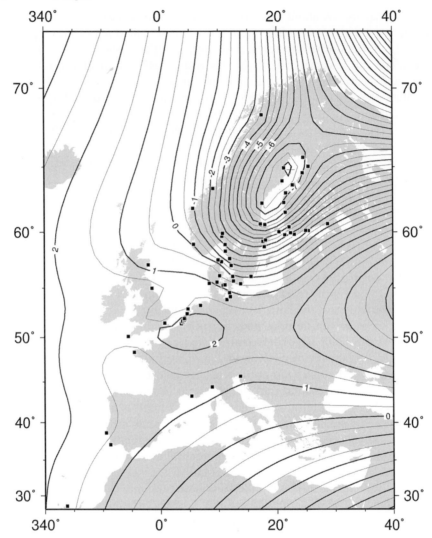

Figure 1.2 Contour plot of 20th-century European sea level trends (mm/yr).

everywhere on the planet in comparison to global estimates of sea level rise and must be accounted for.

Sea level records are also contaminated by local and regional effects such as plate tectonics, extraction of underground fluids (gas, oil, water), or seasonal/ interannual oceanographic effects such as the El Niño phenomenon. All of these have the potential to obscure or hide the long-term global rate of sea level rise.

1.3   CHARACTERISTICS OF SEA LEVEL RECORDS

Climate-related variations of sea level, although small compared to some other geophysical phenomena, share a common property in their spectrum, that is, in the way their power is correlated with frequency. In many geophysical data series large changes are associated with long periods. A few common examples are that small earthquakes are more frequent than large ones, great storms occur less often than small ones, and large meteorites hit the earth less frequently than small ones. Temporally, global or regional sea level varies on a scale of about 100 m over the glaciation–deglaciation cycle, but only a few tens of centimeters due to the seasonal heating cycle of the upper ocean, or a seasonal–interannual El Niño event.

Another characteristic of climate-related sea level variations is that they can be spatially correlated at low frequencies, such as seasonal and longer periods. Figure 1.3 displays smoothed and normalized (i.e., divided by the

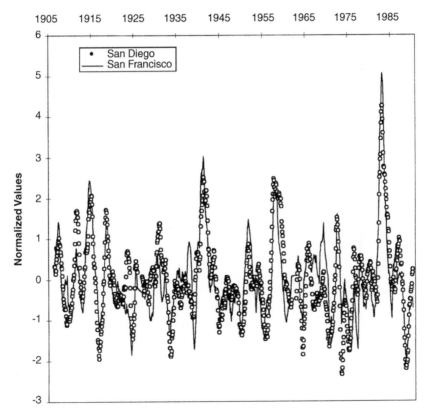

Figure 1.3   San Diego and San Francisco normalized and detrended relative sea levels (mm/yr).

root-mean-square values of the series) monthly mean sea level records for San Francisco and San Diego, California, during their common observation period that began in 1906. (Their trends of about 2 mm/yr have also been removed for the sake of clarity.) The agreement at interannual and longer periods is very striking, even though these two sites are about 800 km apart. Such high spatial correlation at low frequencies is exhibited nearly everywhere and plays a critical role in attempts to determine a global rate of sea level rise from tide gauge data. Merely adding more tide gauge records to a solution for global sea level does not necessarily make any improvement if the additional data records are highly correlated with existing ones and sample only a water mass already sampled. On the west coast of North and South America, the large interannual–interdecadal variations of sea level are caused by El Niño/La Niña events (note in Fig. 1.3 the especially large excursions during the 1982–83 El Niño period), as pointed out in the classic paper by Chelton and Davis (1982). Sturges and Hong in Chapter 7 of this book show that the low-frequency variations of sea level that occur on the east coast of the U.S. have a origin different from those on the west coast, resulting from wind stress on the Atlantic Ocean. It is important to keep in mind that sea level variations are driven phenomena, not drivers, meaning that they are a result of forcings by other geophysical events. The tides are a perfect example of this, since they are an outcome of the varying gravitational forces of the sun and moon on the earth. Going beyond the tides, a time history of sea level, either global or local, is the sum of many contributors, including thermal expansion, melting of glaciers, wind forcing, and currents.

Sea level variations that occurred relatively soon after the peak of the last glacial maximum about 21,000 years ago are very different from those seen today. Sea level rose about 125 m globally by the conclusion of the melting, but with significant regional variations due to viscoelastic readjustment of the earth and modifications of the geoid from redistribution of mass. The concept of eustasy, that is, a uniform change of sea level occurring everywhere from addition or thermal expansion of water, is an inadequate description for the deglaciation event. This is so because the land rebounded by hundreds of meters in the deeply ice-covered regions, land peripheral to the ice sheets subsided, and the seafloor of necessity sank under the weight of the added meltwater. The redistribution of mass also caused a change of the gravitational potential that altered the geoid and hence sea level. As mentioned earlier, even after the melting was complete, the earth continued to adjust from the removal of the load. The complexities of all of these interactions are very great and are the subject of Chapter 4. We can get at least a feel here for the magnitude of the changes of water level due to Holocene isostatic readjustment from the sea level data shown in Fig. 1.4. Displayed are the elevations relative to modern sea level of the Bristol Channel, United Kingdom, site for the last 8 millennia (data adapted from Roberts, 1994, p. 186). In this figure the origin at 0 BP is the present, and sea level

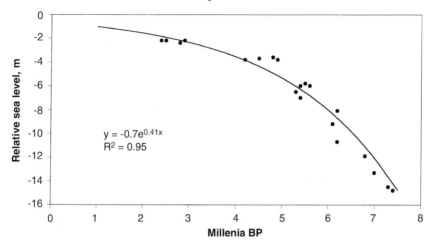

Figure 1.4   Bristol Channel mid and late Holocene relative sea levels (data from Roberts, 1994).

was lower in the past than now. During the time shown in Fig. 1.4, sea level rose in the Bristol Channel area by about 16 m from isostatic readjustment, that is, subsidence of the land, and from changes in water level and the geoid. Note in Fig. 1.4 the nearly exponential nature of the change of sea level. The characteristic decay time (i.e., the time for the level to change by a factor of $1/e$) for this curve is about $1/.41 \approx 2.5$ millennia. The rate of relative sea level rise at 0 BP due to GIA is still about 0.3 m per millennium, or 0.3 mm/yr.

Although eustasy is a poor approximation to sea level change during and immediately after the deglaciation process, it is not far from an accurate description (within perhaps 0.1 mm/yr) for analyses of contemporary sea level rise. It is observed (see Chapter 3) that properly selected long tide gauge records tend toward a narrow range of values for the trend of sea level regardless of location. This merely reflects that the earth does not adjust viscoelastically instantly to the addition of 1 mm or so of water per year to the oceans, and that the purely elastic response is a small fraction of the ultimate long-term viscoelastic adjustment.

The foregoing has indicated that much analysis is required to determine a rate of global sea level rise that can be used scientifically with confidence. A good place to start is with a blunt instrument, in this case, a histogram of sea level trends derived from tide gauge data from around the globe. Trends of relative sea level from records at least 20 years in length were derived from the sea level data base at the Permanent Service for Mean Sea Level in the United Kingdom and grouped (binned) according to value. Figure 1.5 presents the resulting histogram. As might be expected, the scatter of the trends is

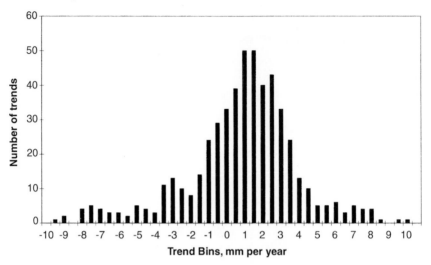

Figure 1.5   Histogram of relative sea level trends derived from individual tide gauge records.

very large and is in addition skewed somewhat in the direction of negative (i.e., falling) values of sea level change. But there is a clear preference for a most probable trend value that is greater than zero. Much of this book will deal with exploring why the range of sea level trend estimates is so broad and which sea level records are appropriate and useful for determining a global trend of sea level.

REFERENCES

Bindschadler, R. (1998). Future of the West Antarctic ice sheet. *Science* **282,** 428–429.

Bruun, P. (1962). Sea-level rise as a cause of shore erosion. *Am. Soc. Civ. Eng. Proc., J. Waterways Harbors Division* **88,** 117–130.

Chelton, D. B., and Davis, R. E. (1982). Monthly mean sea-level variability along the West Coast of North America. *J. Phys. Oceanog.* **12.**

Chovitz, B. H. (1981). Modern geodetic earth reference models. *EOS Trans. AGU* **62,** 65–67.

Cohen, J. E., Small, C., Mellinger, A., Gallup, J., and Sachs, J. (1997). Estimates of coastal populations. *Science* **278,** 1211–1212.

Galgano, F. A., *Geomorphic analysis of modes of shoreline behavior and the influence of tidal inlets on coastal configuration, U.S. East Coast,* Ph.D. dissertation, Department of Geography, University of Maryland, College Park, 1998.

Galgano, F. A., Douglas, B. C., and Leatherman, S. P. (1998). Trends and variability of shoreline position. *J. Coastal Res., Special Issue* **26,** 282–291.

Hands, E. B. (1983). The Great Lakes as a test model for profile responses to sea level changes. In *Handbook of Coastal Processes and Erosion,* edited by P. D. Komar. C.R.C. Press, Boca Raton, FL.

Heiskanen, W., and Moritz, H. (1967). *Physical Geodesy.* Freeman, San Francisco.

Kearney, M. S., and Stevenson, J. C. (1991). Island land loss and marsh vertical accretion rate evidence for historical sea-level changes in Chesapeake Bay. *J. Coastal Res.* **7,** 403–415.

Lambeck, K. (1988). *Geophysical Geodesy.* Oxford Univ. Press, Oxford.

Leatherman, S. P., Zhang, K., and Douglas, B. C. (2000). Sea level rise drives coastal erosion. *EOS Trans. AGU* **81,** 55–57.

National Research Council (NRC) (1990). *Managing Coastal Erosion,* National Academy Press, Washington, DC.

Nicholls, R. J., and Leatherman, S. P. (1994). Global sea-level rise. In Strzepek, K., and Smith J. B. (eds.), *As Climate Changes: Potential Impacts and Implications,* Cambridge Univ. Press, Cambridge.

Nicholls, R. J., and Leatherman, S. P. (1995). The implication of accelerated sea level rise for developing countries: A discussion. *J. Coastal Res., Special Issue* **14,** 303–323.

Pilkey, O. H., and Davis, T. W. (1987). An analysis of coastal recession models: North Carolina coast. In *Sea-level Fluctuations and Coastal Evolution* [Special publication 41], edited by D. Nummedal, O. H. Pilkey, and J. Howard. Society for Sedimentary Geology, Tulsa, OK.

Pugh, D. T., *Tides, Surges, and Mean Sea Level,* Wiley, New York.

Roberts, N. (1994). *The Changing Global Environment.* Blackwell, Cambridge, MA.

Schwartz, M. L. (1965). Laboratory study of sea-level as a cause of shore erosion. *Journal of Geology,* **73,** 528–534.

Schwartz, M. L. (1967). The Bruun theory of sea-level rise as a cause of shore erosion. *Journal of Geology,* **75,** 76–92.

Warrick, R. A., Le Provost, C., Meier, M. F., Oerlemans, J., and Woodworth, P. L. (1996). Changes in sea level, *Climate Change 1995,* Intergovernmental Panel on Climate Change, Cambridge Univ. Press, Cambridge.

Wingham, D. J., Ridout, A. J., Scharroo, R., Arthern, R. J., and Shum, C. K. (1998). Antarctic elevation change from 1992–1996, *Science* **282,** 456–457.

Zhang, K. (1998). *Twentieth Century Storm Activity and Sea Level Rise Along the U.S. East Coast and Their Impact on Shoreline Position,* Ph.D. Dissertation, Department of Geography, University of Maryland, College Park.

Figure 1.1 Geoid variations for the continental United States.

Figure 2.1 Watercolor of an area of modern-day Honolulu by Richard Brydges Beechey painted about 1826. In the lower right-hand corner is a prehistoric fish pond that may be a remnant of a shallow, brackish lagoon thought to have formed during a period of higher than present sea level ca. 3400 to 2200 years ago. This sea level highstand may have been 5.5 feet (~1.6 m) higher than present MSL. Courtesy Peabody Essex Museum, Salem, Massachusetts.

Figure 4.3 Northern hemisphere isopacks of the ICE-4G model of the last deglaciation event of the current ice age for six times beginning at Last Glacial Maximum (LGM) 21,000 calendar years ago and ending at the present. Ice thickness contours [for the "explicit" component of the ice load; see Peltier (1998c) for detailed discussion of the "implicit" component] are drawn in 1-km intervals.

Figure 4.4 Global predicitons of the present-day rate of relative sea level rise delivered by the gravitationally self-consistent theory of postglacial relative sea level change when the ICE-4G model of the LGM to present deglaciation history is employed as input to the calculation. The upper figure is for the VM1 viscosity model, the middle figure is for the VM2 viscosity model, and the lower figure shows the global map of the difference between the predictions based upon these two models of the radial viscosity variation.

Figure 4.5 Global predictions of the present-day rate of relative sea level rise delivered by two versions of the gravitationally self-consistent theory for postglacial relative sea level history, both of which employ the ICE-4G model of the deglaciation history and the VM2 viscosity model. In the calculation that produced the upper figure the influence of rotational feedback was ignored whereas in the middle figure this influence was included. The lower figure displays the difference between these two results, which clearly has the pattern of the degree 2 and order 1 spherical harmonic, demonstrating that the rotational feedback is dominated by the influence of polar wander, the influence of the changing rate of axial rotation playing a minor role.

Figure 4.7 Gravitationally self-consistent model predictions of the rate of absolute sea level change or equivalently the rate of change of geoid height that we expect to be directly observable in the context of the GRACE experiment. The upper and middle components of this figure show the two components of the time rate of change of geoid height field, respectively the rate of RSL rise and the rate of change of the local radius of the solid earth relative to the center of mass. The lower frame shows the sum of these fields for the ICE-4G (VM2) model and thus the time rate of change of absolute sea levels for a calculation in which the influence of rotational feedback has been neglected.

Figure 4.8 Same as for Fig. 4.7 except that the influence of rotational feedback has been included in the calculation.

Figure 4.9 The upper and middle plates show the time rate of change of geoid height fields predicted by the ICE-4G (VM2) model excluding and including, respectively, the influence of rotational feedback. The lower plate shows the difference in these predictions.

Figure 4.12 Present-day rate of RSL rise as function of geographic position for the ICE-4G (VM2) model (upper figure), for the ICE-4G (VM2) model that includes an Antarctic derived "melting tail" of strength 0.25 mm/yr (middle figure) and for ICE-4G (VM2) including an Antarctic derived tail of strength 0.5 mm/yr. It will be noted that when a "tail" of the latter strength is incorporated, the equatorial Pacific highstand of sea level is entirely eliminated as also shown by the predictions for specific locations in Fig. 4.11.

Figure 5.2 Schematic diagram illustrating the effects on sea level of anthropogenic intervention in the hydrologic cycle. R, reservoir impoundment; G, groundwater mining; I, irrigation; U, runoff from urbanization; C, water released by carbon emissions; D, runoff due to deforestation; and W, wetland drainage. Reprinted from *Global and Planetary Change*, vol. 14, Gornitz, V., Rosenzweig, C. and Hillel, D., "Effects of anthropogenic intervention in the land hydrologic cycle on global sea level rise," p. 148, ©1997, with permission from Elsevier Science.

Figure 5.3 Distribution of the world's major dams (after Chao, 1995).

Figure 6.11 Maps of the amplitude and phase of annual and semiannual sea level variations as observed by T/P over 1993–1998.

Figure 6.12 Sea level trends during the T/P mission (1993–1998) (top map) and the same trends for sea surface temperature changes during the T/P mission (bottom map).

Figure 6.13 The first four leading EOFs of sea level and sea surface temperature. The temporal modes have been scaled to represent their contribution to the global mean variations in these quantities as shown in Fig. 6.9.

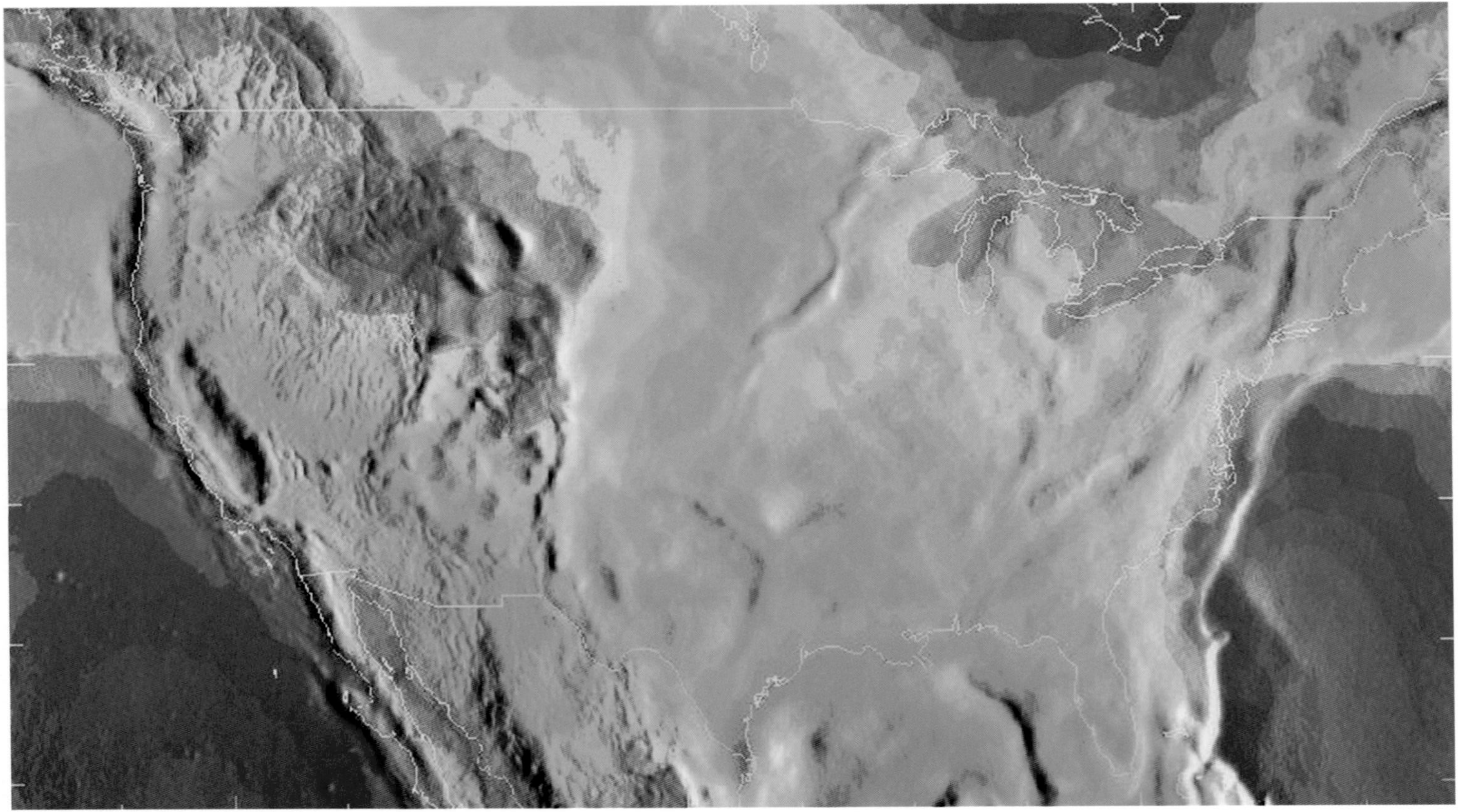

Figure 1.1

Figure 2.1

Ice Thickness

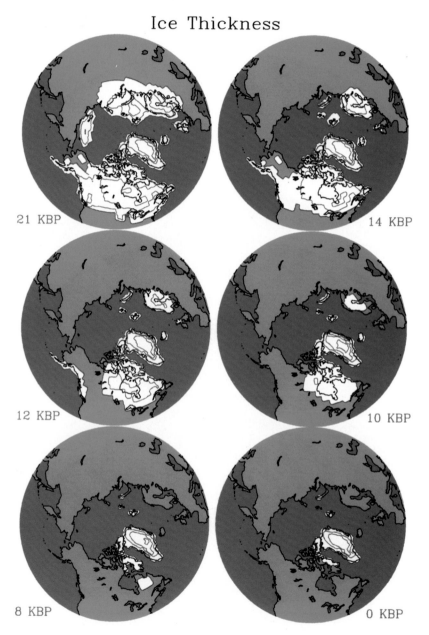

Figure 4.3

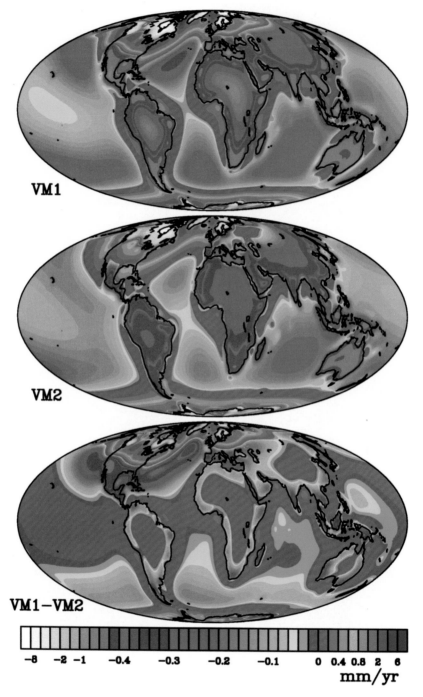

Figure 4.4

Effect of Rotation on
Rate of change of Sealevel

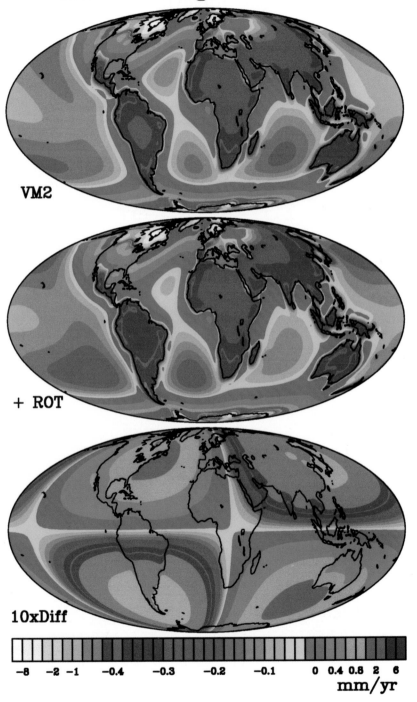

VM2

+ ROT

10xDiff

-8   -2 -1   -0.4   -0.3   -0.2   -0.1   0 0.4 0.8 2 6

mm/yr

Figure 4.5

Rate of Change
(no Rotation)

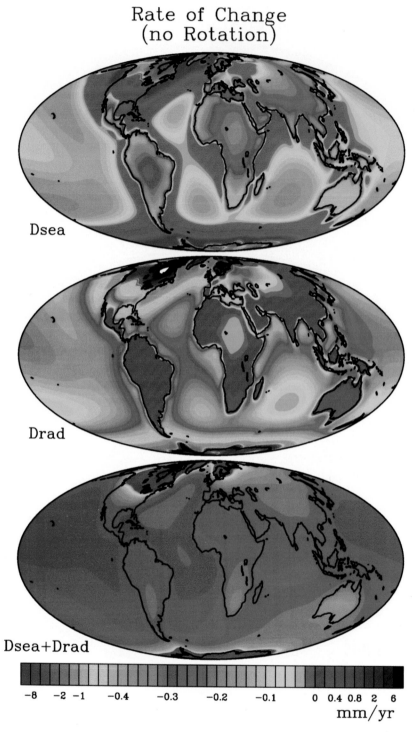

Figure 4.7

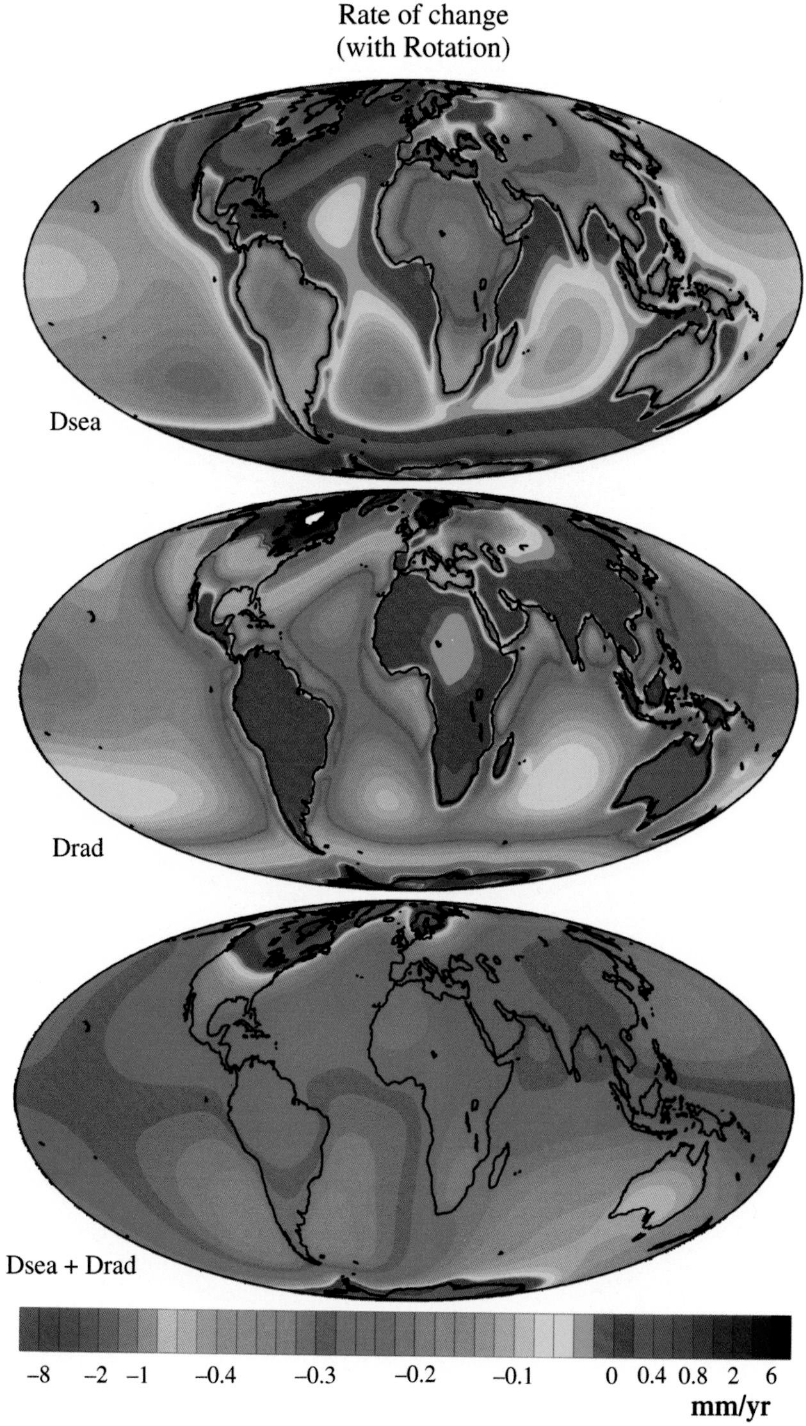

Rate of change
(with Rotation)

Dsea

Drad

Dsea + Drad

-8   -2 -1   -0.4   -0.3   -0.2   -0.1   0 0.4 0.8 2 6
                                                                    mm/yr

Figure 4.8

Rate of Change
of Geoid Height

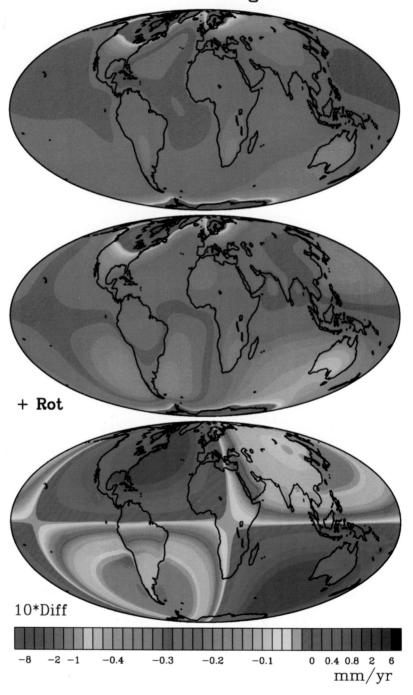

+ Rot

10*Diff

-8   -2 -1   -0.4   -0.3   -0.2   -0.1   0 0.4 0.8 2   6
                                                    mm/yr

Figure 4.9

Present−day rate of RSL variation
predicted by model ICE5G−VM2e−L100

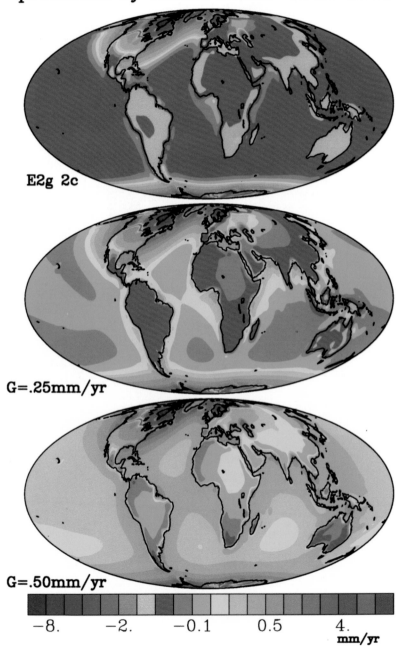

E2g 2c

G=.25mm/yr

G=.50mm/yr

-8.   -2.   -0.1   0.5   4.
                                                mm/yr

Figure 4.12

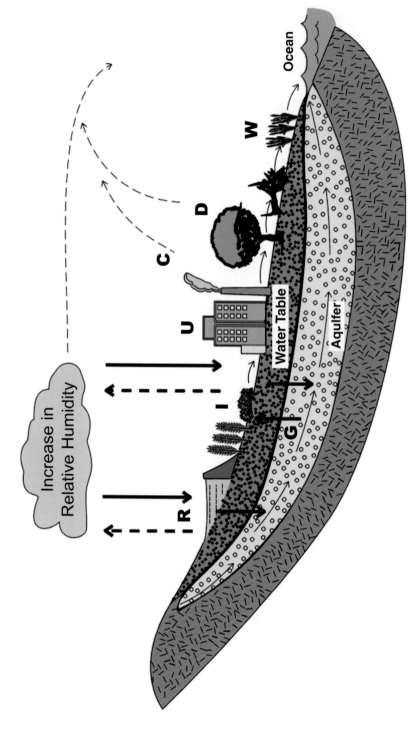

Figure 5.2

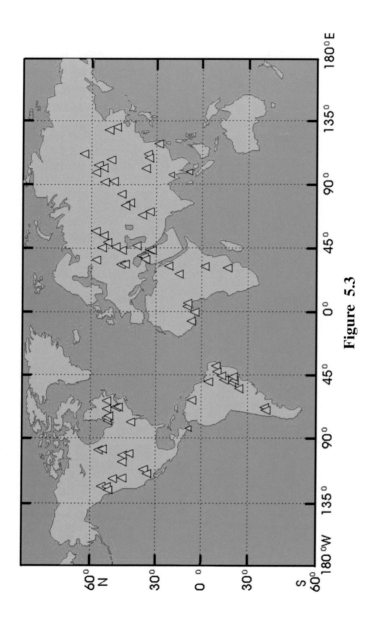

Figure 5.3

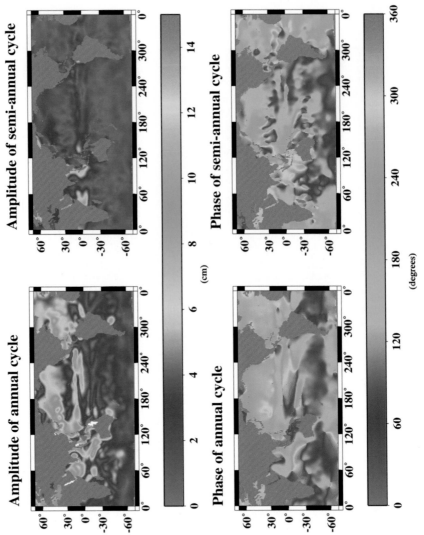

Figure 6.11

Local trends of sea level

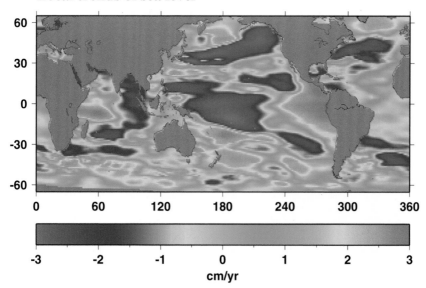

Local trends of sea surface temperature

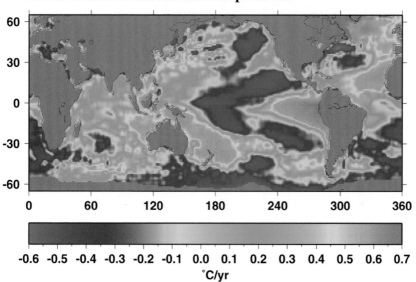

Figure 6.12

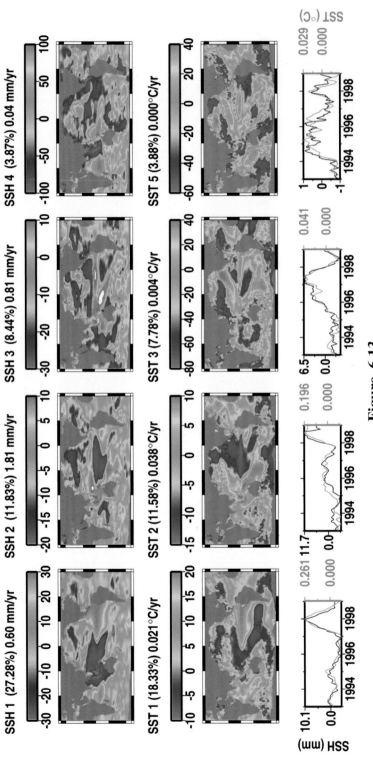

Figure 6.13

Chapter 2 | Late Holocene Sea Level Variation

Michael S. Kearney

> *Is fluctuation possible? The record of fluctuation of glaciers and the curves of fluctuating climate through the last 15,000 years at least admit the possibility of corresponding fluctuations in sea level.*
>
> —R. F. FLINT (1971)

2.1 INTRODUCTION

Almost 30 years have passed since Richard Foster Flint posed and tentatively answered this question concerning the sea level record since retreat of the late Wisconsin ice sheets (ca. 12,000 BP).[1] The breadth of work on late Quaternary (ca. 12,000 BP–present) sea level change (let alone allied glacial and climate variations) undertaken since Flint's query both in detail and in geographic coverage has been prodigious as a glance at the bibliographies of such synthesis volumes by Tooley and Shennan (1987) and Pirazolli (1991) testifies. At present, upward of 500 late Holocene sea level curves exist for various coastal areas worldwide (see Pirazolli, 1991), and the number of dated sea level indicators easily reaches into the thousands. The existing data base of late Holocene sea level change is something Daly (1934), the first to hypothesize the depth of the regression during the last glacial maximum, could only imagine.

The dramatic rise in global sea levels by more than 100 m since the last glacial maximum (ca. 18,000 BP) is universally accepted today. The evidence for this change has been so well described for so many areas (cf. Pirazolli, 1991), and the geophysical models supporting it so persuasive, that a newcomer to the field could be forgiven for thinking there was little left to discover. Nonetheless, in many respects some of the basic questions remain regarding late Holocene sea levels that first intrigued researchers over 30 years ago. The case for sea level fluctuations is far from settled, although the argument today centers less on whether sea level variations actually occurred and more on the reliability of the evidence for them. Even the notion of generally higher sea levels than present during the middle Holocene warm period (ca. 8000–5000 BP) is still discounted by some researchers. In this chapter, the U.S. Atlantic Coast will be used to trace the history of sea level rise during the late Holocene, focusing on the record of the last several thousand years

---

[1] BP means radiocarbon years "before present," present being 1950 AD.

as an example of the relations between climate and sea level. But first, since the nature and quality of the evidence for past sea level change have often informed the discussion of how sea level varied in the past, it is worthwhile to review how the evidence was gathered.

2.2 EVIDENCE FOR PAST SEA LEVELS: THE DATA AND THE CONCLUSIONS DRAWN

In the first decades of the last century, Charpentier and others, examining the large boulders of foreign rocks in major valleys in the Alps, were eventually persuaded that the local peasants were right. Glaciers, then small and restricted to the upper parts of the valleys, had once been far more extensive and carried with them these boulders from source areas farther up the valleys (Flint, 1971). Once convinced that only ice in the form of large glaciers (rather than some preternatural floods) could move such large debris over wide areas, it was inevitable that other large boulders of foreign rocks scattered elsewhere throughout northern Europe would be interpreted as having been transported the same way, only in this case by continental size ice masses. Thus, the Glacial Theory was born.

No isolated musing, however informed, could have arrived at such a conclusion. If the boulders had been next to a large river, it is possible to envision a totally different conclusion arising from these early observations, and establishment of the Glacial Theory delayed for some time. Unfortunately, the historical development of ideas of sea level variations during the late Holocene (ca. 5 BP–present) cannot point to a similar single instance where long, if not widely, recognized physical evidence finally connected with an enlightened reception. However, the type and quality of the evidence have been equally important in how concepts have evolved, and why controversies on the most basic of questions concerning past sea level change remain current.

2.2.1 Geomorphic Features

Historically, geomorphic features have provided the clearest evidence of a former change in sea level. Erosion and inundation during transgressions, or prolonged periods of rising sea level, often erase or submerge former sea level positions. Until the advent of offshore seismic stratigraphy and bottom profiling, the features most often available as possible indicators of former changes in sea level were likely to have been supertidal in elevation (e.g., raised beaches) and landward of existing shoreline position, and thus, representative of a time when sea levels were higher than present. Such evidence was fundamentally at best a very incomplete record of the sea level history of most areas. The actual possibility of higher sea level than at present during the late Holocene is a separate issue that will be discussed later.

In more recent decades, the incompleteness of the record has been seen as less of a problem than whether such evidence documents a change in sea level at all. That raised beaches, wave-cut scarps, and strandlines demonstrate that water levels were at one time higher has not been the issue; rather, were the episodes of former high water levels related to sea level changes, and not storms (Fig. 2.1, see color plate) (see Stapor, 1973)? It has been shown that storm surges can cause either deposition or erosion considerably landward of modern shorelines and, in fact, raised beach ridges have been use to reconstruct the tempo of storm periodicity since the early Holocene (cf. Fairbridge, 1987). Tsunamis are an additional complication. Although passive continental margins like the U.S. Atlantic Coast are hardly areas where such events can be expected, no coast is probably immune from tsunamis. About 7000 years ago, a submarine slide off Norway (part of the Amero-type trailing continental margin of Northern Europe) generated a tsunami which deposited a sand layer in eastern Scotland almost three-quarters of a meter thick at some sites (Long *et al.,* 1989).

Still, though some remain suspicious about the connection between sea level and old supertidal "shoreline" features, the wide geographic extent of some scarps, like the 1.5-m above-mean-sea-level (MSL) scarp ringing much of the Florida Panhandle, has been argued as something only a transgression could create (Donoghue and Tanner, 1992). Even widespread scarps generated by storms, such as those formed by the 1991–1992 northeasters along the U.S. middle and north Atlantic Coasts, would be unlikely to be continuous and at relatively the same elevation, since local coastal configuration and regional variations in the storms would combine to produce irregularities in surge height and penetration. Perhaps the skepticism that greets reports of raised beaches and former sea level highstands more than anything crystallizes the debate alluded to by Flint in the quote that begins this chapter; highstands are undeniably fluctuations.

2.2.2   Transgressive Sequences

Worldwide the history of Holocene sea level change is written in the sediments left as a trace of its progress across continental shelves. The nature of these sequences varies by coastal type and locale, and the breadth of deposits and their possible interpretation would easily require a separate chapter. However, whatever the coast, transgressive sequences are generally the only way information can be obtained on sea level changes predating the last several millennia, when sea levels often stood many meters below modern limits on what today are inner continental shelves.

The transgressive stratigraphy of the barrier island coasts of the U.S. Atlantic and Gulf Coasts can serve as an example of the potentials and liabilities of these data sources. In these systems, sandy barriers—such as the long thin barriers like Atlantic City, New Jersey, and Assateague Island, Maryland—

and their associated lagoons and marshes migrated tens of kilometers across the inner continental shelf (Fig. 2.2). The lagoonal marsh deposits are the keys to deciphering these transgressive sequences, as they provide the medium for radiocarbon dating, and there has been considerable discussion concerning their preservation potential during a transgression. Depending on wave energy and the resulting depth of shoreface erosion, these materials either may be preserved intact or may suffer total removal (Kraft, 1971). If the latter occurs, the only vestiges of the transgression are nearshore marine sands, left after beveling of the shoreface by erosion.

According to the model proposed by Belknap and Kraft (1981) and Kraft and Chrzastowki (1985), the usefulness of a coastal transgressive record is linked to its preservation potential. If we assume a relatively smooth, nondissected preexisting topography where transgressive facies cannot hide in antecedent valleys, extensive preservation of barrier and lagoonal facies is indicative of rapid sea level rise where quick submergence isolates these materials from wave erosion. Alternatively, loss of the transgressive facies suggests belated submergence below wave base and slower rates of sea level rise. In theory, therefore, a sea level record punctuated by periods of slow and rapid rise could be gleaned from the transgressive sequence. Ideally, also, a true picture of former sea level elevation could be obtained. In the original Fischer (1961) model, landward migration of a barrier progressively buries the peats of marshes fringing the transgressing mainland shoreline of the lagoon. Dating

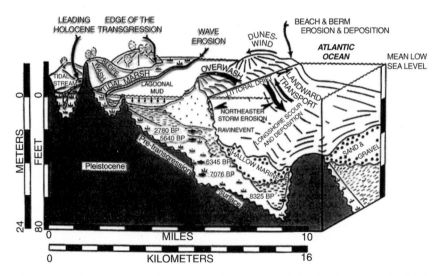

Figure 2.2 Transgressive stratigraphy of the Delaware coast, showing the present-day barrier island and seaward sand sheet overlying older lagoonal materials. Radiocarbon dates are on lagoonal materials (basal peats, shells, and wood) which can be used to estimate relative rates of landward shift of the transgressive barrier-lagoon coast as sea levels rose (from Kraft *et al.,* 1973).

the peats at the contact of the pretransgression surface of the shore profile could furnish a dependable means of securing reliable information on sea level position, similar to the spreading of coastal marshes across upland surfaces with rising sea level described later. Nevertheless, existence of a continuous peat layer seaward of modern barriers has not been reported for many areas. More often the peats are spotty in occurrence, localized in old valley systems (Riggs *et al.*, 1992), and most importantly, not in irrefutable contact with pretransgression surfaces. In the Virginia barrier islands, Oertel *et al.* (1992) found the pre-Holocene (i.e., pretransgression surface) within a few meters of the contact of the sand prism of the modern barriers, with no evident buried peats. Seaward of these islands, the shoreface appears to consist mostly of muds—no convincing transgressive stratigraphy, and little material to date.

The impact of a fall in sea level, or regression, in barrier coasts theoretically can result in progradation or seaward growth of the system, creating a series of beach ridges if sediment supply is adequate. Marshes often develop in the swales between the ridges, and by dating the basal peats in these marshes age limits for the ridges can be obtained. But interpreting these complex morpho-stratigraphic features with respect to sea level history can be daunting, as "regressive" barriers like Kiawah Island, South Carolina, or Galveston Island, Texas, have evolved during the overall transgression of the last several millennia. High sediment supply, in each case for different reasons, simply overwhelmed sea-level–driven shoreface processes that drive barrier retreat, and the barriers prograded.

The terminological morass of "regressive–transgressive" barriers—prograding barriers in the face of rising sea levels—is one that continues to confuse the unwary into interpretative cul de sacs. The problems are certainly illustrative of the pitfalls transgressive/regressive sequences pose for elucidating the sea level history of an area. Barrier transgressive sequences are demonstrably the result of prolonged sea level rise, but the precision with which the transgression that produced them can be deduced is wholly dependent on the preservation of lagoonal peats that provide dating control. This, in turn, is a function of the wave power, antecedent geology, and sediment supply, all of which may or may not have operated in ideal combination in any one area. By comparison, regressive sequences are not demonstrably the result of a prolonged regression, or fall in sea level, and in older shelf deposits, absent other supporting evidence, they can lead to erroneous conclusions regarding past regressions that never occurred.

2.2.3   Biological Indicators of Sea Level Change

Biological indicators, like shells or tree stumps, for reconstructing former sea levels can be very persuasive evidence of significant sea level changes. This is the case especially where fossils of upland tree species are recovered from considerable depth of water in continental shelves or where fossils of deepwater fauna crop out in sediments occurring substantially above modern sea level. One need only recall the example of Charles Darwin, clambering among rocks high in the Andes during the voyage of the *Beagle,* and the inescapable conclusions he reached regarding crustal uplift upon encountering fossils of rugose corals. Similarly, fossils of mammoths and mastodons found in surficial sediments of the inner shelf of the U.S. east coast (Emery, 1967) testify to the depth of the regression during the last glacial maximum.

Unfortunately, as telling as this evidence can be, there are numerous problems that confront any researcher in attempting to infer more than relative changes in sea level position from such material. The first question that must be answered is, was the fossil organism buried in place (i.e., *in situ*) after death? Rooted tree stumps are an obvious instance where it is generally safe to assume that no post-mortem transport of the organism occurred (Fig. 2.3). However, with logs, for example, this can be a risky assumption—the author once found a large remnant of a telephone pole lodged in the beach sands of Parramore Island, Virginia, with a stamp indicating it had originated in New Jersey some 300 miles north! Shells of smaller marine organisms like oysters and clams can also be transported considerable distances upon death, depending on currents and the incidence and intensity of coastal storms. Ultimately, any researcher must determine whether the organism remains are in life position or, in the case of shell material, a life assemblage. There is a considerable paleontological literature on this (cf. Easton, 1960). But if the material lacks obvious signs of transport like abrasion or breakage—not uncommon for benthic species unexposed to wave action and significant transport—it may be very difficult to judge whether it is *in situ.*

A second, and perhaps more meaningful, question to ask is, what is the relation of the species' habitat to mean sea level—how accurately can you reconstruct former sea level position upon finding fossils of coastal marine or mainland species in buried littoral sediments? Kidson and Heyworth (1979) and van der Plassche (1977), among others, have reviewed extensively the errors in determining past sea levels from material as diverse as tree stumps to the shells of shallow water bivalves. As any inveterate beachcomber knows, many pelecypod species can occur over a wide range of depths, often from the surf zone down to more than 10 m depth below mean sea level (Fig. 2.4). The example of *Mytilus edulis* is illustrative. In German Bay in the North Sea, this species has been reported to occur from the shelf bottom (depth of water around 20 m) to relatively shallow channel areas (<5 m depth) (Reineck and Singh, 1975). Hence, the best that can be said for fossils of *Mytilus* is that they have only "a crude relationship to sea level" (Kidson, 1982), and incorporating these data in sea level records requires a realistic appraisal of possible errors. As Kidson (1982) noted, the case for past sea level fluctuations is hardly unimpeachable with such sea level indicators.

Submerged tree stumps of riparian species have figured prominently in many sea level records. The danger here is similar to injudicious use of pelecy-

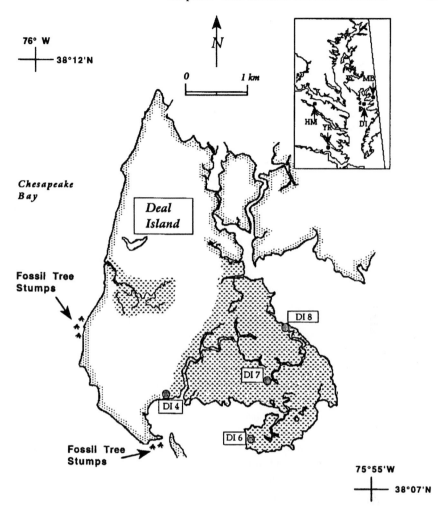

Figure 2.3   Use of rooted tree stumps occurring along Deal Island, Maryland, for reconstruction of sea level changes during the last thousand years in the region. The tree stumps, dating from around 800 BP, were exhumed from the overlying nearshore sands by a storm. In dating such material, there are several problems, not the least of which is contamination of the old wood by boring marine organisms. A more subtle problem is the actual age of the tree upon death, which relates to the timing of sea level change. It is highly possible that these trees were dead snags long before burial by the sand layer, probably being killed by inundation of their roots by an increase in groundwater level as sea level first rose. Thus, the age of the trees may predate actual submergence of the site. (Reproduced from Kearney, M. S., Sea level change during the last thousand years in Chesapeake Bay. *Journal of Coastal Research,* **12,** 977–983, 1996.)

pod shells, since many riparian tree species are only nominally so, and can survive quite easily in upland environments. An example is loblolly pine (*Pinus taeda*), a common tree in shoreline areas where soils are poor along the

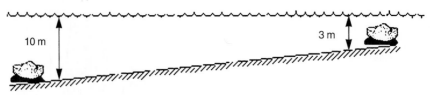

Figure 2.4   Diagram illustrating the potential depth variation in the occurrence of a common pelecypod species, such as *Mytilus edulis*.

U.S. Atlantic Coastal Plain—even in high levees of Chesapeake Bay coastal marshes—but ranges into the Piedmont (Silberhorn, 1982). Tree species like bald cypress (*Taxodium distichum*) are at least restricted to aquatic, if not necessarily shoreline, habitats, but do not occur in marine environments, tolerating little salt. Thus, finding *in situ* fossil wood of a species like bald cypress buried beneath littoral sediments certainly is powerfully evocative of a dramatic rise in sea level; however, not knowing where the tree was growing in relation to past shorelines before submergence and death obviates any real reconstruction of the degree of shoreline change.

   Ideally, the most useful biological indicators of sea level change are those whose occurrence can be demonstrated to be restricted to within a very narrow elevation/depth range with respect to mean sea level—Kidson (1982) suggests about ±1 m. Even that restricted a range is still too coarse for determining sea level changes within the last 2 millennia, where the total change in mean sea level in most areas has been only on the order of a few meters. Of course, like all biological indicators of past environmental conditions, the fossil must be common in older sediments and resistant to problems of postburial contamination or changes in chemical composition which introduce errors in the typical radiometric dating techniques (e.g., ^{14}C dating) for determining age. Regrettably, biological sea level indicators used in many sea level records often fall short in meeting one or more of these criteria.

2.2.4   Archaeological Data

People have always lived along the coast, as a source of food and transportation. This trend continues today, with 100 million people worldwide presently living within 1 m of mean tide level (Houghton *et al.*, 1996). As sea levels rose during the Holocene, inundation of settlements and structures from the late Neolithic onward has provided a wealth of potential information for the sea level researcher. The advantages of archaeological data are clear enough. First, dating, particularly with classical cultures of the Mediterranean, Mesoamerica, and elsewhere, can be very precise, often to within decades using pottery and other artifacts. Second, the relations between the evidence and mean sea level, most notably structures from ancient ports, can also be drawn fairly precisely. A recent archetypal use of archaeological data for reconstruction of sea level history is the work undertaken in the Severn Estuary of southern Britain, where Neolithic trackways, Bronze Age waterlogged wooden buildings, Roman military ditches, and Medieval dikes in coastal wetlands and nearshore tidal flats provide detailed evidence of transgressive and regressive episodes dating back some 3000 years (Allen, 1987; Allen and Rae, 1987; Allen and Fulford, 1988; Whittle, 1989).

Apart from Spanish colonial-era structures, like remnants of an early 17th-century fort at Parris Island, South Carolina, that lies below MSL today (National Academy of Sciences, 1987), shell and other middens are the major archaeological archive for sea level information for the U.S. Atlantic Coast prior to British settlement. Indeed, much of what has been proposed for the late sea level changes in South Carolina and Georgia (Colquhoun and Brooks, 1986; DePratter and Howard, 1977) has relied on middens and associated artifacts. However, few data on past sea levels have been viewed with more suspicion. The rationale is actually quite reasonable: native peoples, having learned to their travail of the great perishability of shellfish, almost immediately cleaned and consumed their catches at the handiest location, the shoreline. Naturally, the sites preferred were those also close to proven sources of shellfish, like oyster reefs. Over time considerable mounds of shell debris accumulated, creating the midden (see Fairbridge, 1976).

In fact, there are ample grounds for believing that this is generally what happened. Most buried shell middens reported in the literature are often in contact with intertidal sediments like salt or estuarine marsh peats (e.g., Colquhoun and Brooks, 1986) or beach sands and ridges (e.g., Stapor *et al.*, 1991). Thus, middens within a carefully worked-out stratigraphic/geomorphic framework can be considered shoreline markers of a sort, but the debate centers in how close to the former shoreline and how high above the past sea level position. Fairbridge (1976) described criteria for the identification of paleo-Indian campsites in Brazil and suggested that most sites were established at least 1 m above normal high tide to prevent flooding even by exceptional events like storms. However, the top of the midden can rise over time substantially above MSL; Fairbridge (1976), in studies of Brazilian middens, suggests as much as 20 m until finally abandoned. Hence, very large middens, built within the last several thousand years, can only provide sea level elevation minima, since apart from postestablishment erosion or destruction, they are unlikely to have been inundated and buried (cf. Fairbridge, 1976; Fig. 2.5).

As testaments to human interactions with the coast, archaeological data add an appealing dimension to sea level studies that perhaps begs the issue of their suitability in many instances. But, in the best-case scenarios where their relationships to former sea levels can be reliably determined, they can provide a measure of dating control (often down to within a decade) that exceeds most other methods.

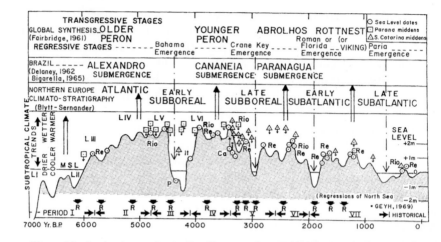

Figure 2.5 Sea level curve for the Brazilian coast from Fairbridge (1976). Note the inclusion of midden dates in the construction of the curve, which are generally construed as indicating transgressive phases.

2.2.5 Intertidal Deposits: Coastal Marshes

The recognition of coastal marshes as repositories of information on past sea levels goes back at least to Johnson's (1925) insight that only global sea level rise, not just the vagaries of local land subsidence, could explain the worldwide occurrence of coastal marshes (cf. Newman *et al.*, 1980). This, together with earlier observations of Mudge (1858)—who asked the question, "whence all this mud? . . . ."—and Shaler (1885) that marshes grow upward and laterally (vertical and lateral accretion) by trapping mineral sediment carried in with the tides, has led to the increasing investigation of marsh sediments as the preferred archives of regional sea level history (Stevenson *et al.*, 1986). At present, coastal marshes are being investigated for evidence of regional changes in sea level by dating the sediments themselves and by examining faunal remains and geochemical indicators of former sea level position.

Modern study of marsh deposits for reconstructing past regional variations in sea level began after World War II. The first sea level curves based on marsh records that are still cited today began to appear in the 1960s (e.g., Bloom and Stuiver, 1963). After Kaye and Barghoorn (1964) published their seminal paper on the autocompaction of older marsh peats, it was quickly recognized that a single marsh sediment core would no longer suffice for an accurate depiction of changes in sea level elevation over time. Essentially, because older peats collapse under their own weight, as well as lose volume from decay and dewatering, they tend to sink lower than the depth at which they formed in relation to a former sea level. The amount of collapse and subsidence of old marsh peats increases with age (i.e., depth), and although this could conceivably be accommodated by developing a functional model of age versus autocompaction, it would have to be assumed that the lithology of the peat (especially bulk density) remain unchanged over time. Unfortunately, this is decidedly not the case, as the development of the marsh produces changes in peat composition in response to temporal shifts in plants species across the marsh and the differences in sediment trapping they cause (cf. Ward *et al.,* 1998). Moreover, these considerable obstacles to deriving any relationship of autocompaction to age can be further complicated by the occasional large quantities of mineral sediment that overwash the marsh during storms (Stumpf, 1983).

It was realized some time ago (cf. Redfield, 1972) that a reliable record of vertical growth of the marsh as a result of rising sea levels could be obtained by analyzing the horizontal spread of the marsh (lateral accretion) across mainland surfaces as they are submerged by sea level rise. In this approach, only the basal peat in contact with the old upland surface is dated. Since only thin wedges of peat at the highly compacted, old surface are dated, the problem of autocompaction is effectively solved, and a true picture of former sea level elevations is possible (Fig. 2.6).

Of course, there are always additional problems. The most important is being aware of just what part of the marsh formed on the upland surface as it was being submerged. Surface elevations change across marshes, generally rising from the shoreline to the upland boundary. In true Atlantic Coast salt marshes, this gradual elevation has been divided into principal zones: the low marsh dominated by a single species, *Spartina alterniflora,* and the high marsh, again dominated by single species, *Spartina patens.* At the upland margin, the marsh surface may be flooded only during the highest mean water conditions (i.e., spring tides), or what is referred to as the mean high high water (MHHW) mark. The connection of this upper landward boundary with the tidal exchange that influences most of the marsh is thus less direct and, by extension, also with mean sea level. This also theoretically comprises the leading edge of the marsh as it first colonizes the newly drowned upland surface. To further

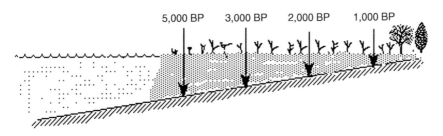

Figure 2.6 Sampling strategy for reconstruction of Holocene sea level rise using the encroachment of the marsh upon the former upland surface (lateral accretion) as sea levels rose.

confuse matters, the plants that characterize this part of the marsh, like *Juncus* sp. or *Phragmites* sp., can also be found in essentially freshwater environments, with little connection to mean sea level (Fig. 2.7). In salt marshes of the open coast, where the sea was presumably somewhere nearby throughout the history of the marsh, at least it can be assumed that the basal peat (even if consisting of upper marsh boundary species) does indicate in some way the first intrusion of a rising sea upon the land. In estuaries, however, nontidal, riverine marshes probably came slowly under the influence of an advancing saltwater front driven by coastal submergence. Here, basal peats may only record the establishment of the marsh, but not sea level. The actual transition to a truly brackish, estuarine marsh may have occurred long afterwards and this, of

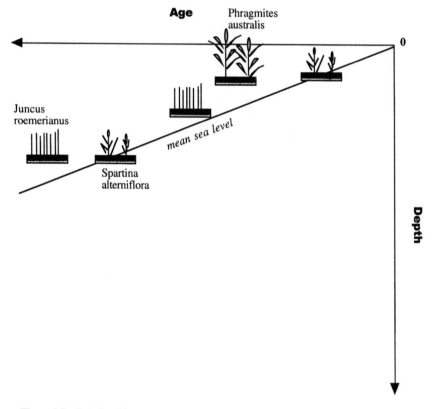

Figure 2.7   Relationship of various common salt marsh plants to former mean sea level. As indicated, the actual elevation of the plants to MSL can vary significantly, perhaps as much as 1 m in some marshes. While the potential error is probably negligible to the determination of former sea level position in the early (ca. 10–8 BP) and middle (8–5 BP) Holocene, when relative sea levels were tens of meters below modern limits, it becomes critical in the reconstruction of the sea level record of the past several millennia where the total range of sea level change is often only a few meters.

course, would have been farther up the sediment column in levels affected by autocompaction.

Scott and Medioli (1978) pioneered the use of marsh foraminifera to delineate former zones in marshes as a means of obtaining better insights into how sea level variations are mediated through changes in marsh environments. Together with closer attention to marsh peat composition (Allen, 1977), marsh sediments can be repositories for the fine-scale records of sea level change requisite for the past several thousand years. Moreover, new approaches, like those of Varekamp *et al.* (1992), demonstrate that the possibilities for extraction of sea level information from salt marsh sediments are still not fully exploited. Geochemical and other parameters tied to changing water levels in marshes can illuminate a richer record of past sea level history than is possible in the general road map of former sea level positions demarcated by basal peat curves.

This last point highlights the intrinsic limitation of basal peat curves. Regressions, even still stands, are not well served by a sole reliance on basal peats because fundamentally such curves are predisposed to portray rising sea trends as the marsh upland boundary extends landwards with further submergence. By comparison, a fall in sea level, especially if prolonged, lowers water levels in the marsh and can lead to oxidation and degradation of surface sediments— hence loss of existing record. Thus, consider the old cliché that the child is father to the man, and whether the (alleged) smooth, nonoscillating sea level curve has been truly independent of the type of data that created it.

2.3   THE HISTORICAL RECORD AS EVIDENCE FOR SEA LEVEL VARIATION

The late 17th century witnessed some of the coldest years since the end of last glacial period. Though this period also saw the dawn of modern instrumental records for changes in climate (Gribbin and Lamb, 1978), probably the most persuasive evidence for the severity of the winters at this time comes from the ample historical documentation. The famous scenes of frozen canals in Holland by contemporary Flemish painters are perhaps the most dramatic examples of winters seldom encountered since in the Netherlands.

The coincidence of extensive historical documentation with the occurrence of the most significant cold period of the late Holocene represents a unique opportunity for climatic research that yields dividends of human interactions to past climates beyond an accounting of temperature. Investigation of sea level variations using historical documents could prove equally rewarding, particularly within the period of the past several centuries which preceded modern mareograph records, but which is still too recent for conventional radiocarbon dating. Woodworth (1999) has used tidal heights recorded in port facilities at Liverpool in the Mersey Estuary since the 18th century to reconstruct sea level variations in southern Britain prior to first tide gauge records of the area. These data indicate a sharp rise in sea levels after the middle of the last century, echoing Horner's (1972) earlier findings of indication of rapidly increasing water levels at London Bridge after the 1780s. Both reconstructions for sea level changes in the Thames are in general agreement with the picture of an upward trend in sea level after about 1820 in the well-known Amsterdam tide staff record (Morner, 1973).

In North America, Kearney and Stevenson (1991) used evidence of land loss in islands of the Chesapeake Bay documented in probate, wills, and other historical records dating from the mid-17th century to infer changes in sea level prior to the period of modern tide gauge records. However, there have yet to be published direct measurements of changes in water level in harbors and estuaries similar to those from Europe. Most likely, the absence of such evidence reflects the large-scale development of North American ports during the 19th century, with demolition or burial of the older colonial structures. For example, remnants of the harbor of New York at the time of its founding as New Amsterdam by the Dutch are occasionally exhumed in the process of new construction, generally covered by landfill and the remains of 19th century structures. A similar story characterizes Baltimore, founded in 1705, whose port really experienced its major growth in the 19th century.

Nevertheless, in North America historical evidence of sea level change in recent centuries has yet to be fully exploited. There continues to be a pressing need for information on sea level changes between the late Middle Ages (~late 15th century) and the 18th century, a time of dramatic change in global climate, as temperatures worldwide began to plunge into what may have been the coldest period in the last 5000 years, the Little Ice Age. The circumpolar expansion in middle and high latitude glaciers in the Northern Hemisphere (and the Andes in the Southern Hemisphere) (Grove, 1988) argues for a comparable-scale oceanic response. But traditional geological approaches, hampered by the limitations of radiocarbon dating involving materials <500 years old and the problems of mixing and contamination of recent sediments, will not supply the answer. At the same time, the application of historical records to the reconstruction of environmental events is fraught with dangers, especially in the instances of largely subjective accounts. Seldom will there be an exact accounting of water level variations, unless structures affected by such changes are still extant.

2.4   THE LATE HOLOCENE SEA LEVEL RECORD

In 1961 Rhodes Fairbridge published what has become the most widely known curve for global changes in middle to late Holocene sea levels (Fairbridge, 1961; Fig. 2.5). Though Fairbridge long ago acknowledged the "primitive" character of early versions of the curve (Fairbridge, 1995), which led to a number of revisions (cf. Fairbridge, 1976), it is still the notion of oscillating sea levels so boldly presented in his original curve that sparks debate. As recently as the late 1980s, the "straw man" of the 1961 curve was argued in juxtaposition to the portrayal of a smooth, continuous rise in late Holocene sea levels to which most sea level data for the U.S. Atlantic Coast appear to conform (Cronin, 1987).

However, in recent years the argument of smooth versus fluctuating sea levels may have become passé. How could it be otherwise, with gathering evidence for late Holocene sea level variation documented in new approaches not available a decade ago (Varekamp *et al.*, 1992; Kearney and Stevenson, 1991); new, convincing support for the concept of cycles in late Holocene climate (Finkl, 1995); and the growing appreciation of the propensity of the "climate machine" to abrupt shifts in state? In spite of all this, it is instructive to examine how these differing views of late Holocene sea level change came about, and what they mean for understanding the relations between climate and sea level.

2.4.1   Smooth (Nonoscillating) Sea Level Curves or "Shepard Curves"

In the 1960s through the 1980s, the smooth, nonoscillating sea level curve came to be viewed as the preferred portrayal of sea level change during the late Holocene. It is not surprising that this occurred, as evidence on the potential errors of many sea level indicators was amassed and the pitfalls of radiocarbon dating became more widely known. Recognition of the latter, of course, changed forever the notion of radiocarbon dating as a precise calendar of past events and led to increased efforts to control problems of contamination with foreign organic materials.

Thus, if any sea level indicator was probably suspect, with regard to both its accuracy in recording former sea level position and its age, the notion of "connecting-the dots" for constructing sea level curves stood largely discredited. The best that could done was to display the source of errors as explicitly as possible (using error bars, or similar devices, for depth and age) and to determine the relative trend in sea level over time. This guiding principle (if it can be called such) of sea level research along the U.S. Atlantic and Gulf Coasts for the past 30 years was not shared by European workers, who, despite admitting the errors in time and sea level position inherent in the construction of sea level curves, continued to investigate the possibility of sea level fluctuations (cf. Shennan, 1987).

The classic sea level curve of Kraft and his co-workers (Belknap and Kraft, 1977; Kraft *et al.*, 1987) for the Delaware coast is one of the best and most widely cited smooth sea level curves for the U.S. Atlantic Coast (Fig. 2.8). Though the curve is mainly based on 16 dated basal peats, there are also over 70 additional dated materials (nonbasal marsh peats) incorporated in the overall sea level record of the area. Because the curve spans almost 12,000

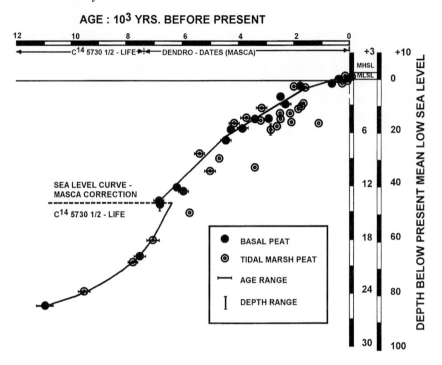

AGE : 10³ YRS. BEFORE PRESENT

Figure 2.8 The classic Holocene sea level curve for the Delaware coast of Kraft and his co-workers. This particular curve is an updated version based on ¹⁴C dates published in Kraft (1976), Belknap and Kraft (1997), and Chraztowski (1986). Figure supplied courtesy of Daria Nikitina and James Pizzuto, Department of Geology, University of Delaware.

years of sea level history (and total changes in sea level on the order of 30 m), the oldest dated basal peats are of necessity not from existing coastal marshes, as the rise in sea level during the mid-Holocene appears to have been too rapid for long-term marsh survival (Rampino and Sanders, 1981).

The curve shows the steady diminution in the rate of sea level rise so typical of late Holocene sea level curves. Submergence of the Delaware coast appears to have been most rapid prior to 5000 BP, when sea levels rose at an average rate of 3 mm/yr (30 cm per century). Interestingly, this figure is considerably slower than the 10 mm/yr (1 m per century) that is conventionally cited as the rate of mid-Holocene submergence of most of the U.S. Atlantic Coast. During the past several millennia, sea levels along the Delaware coast rose at a rate of about 13 cm per century, although there are few dates in the original curve younger than 2000 BP. There is, however, a suggestion of a considerably slowing in the rate of sea level rise after 2000 BP.

More recently, Varekamp *et al.* (1992) discussed the climate–sea level linkages for the last 1500 years for the Connecticut coast, generating a quasi sea level curve based on benthic foraminiferal assemblages and changes in iron abundance. This curve, though not a classic basal peat curve, does suggest a distinct oscillating character to the sea level record of the last millennium and a half, with distinct regressions and transgressions, and obvious decelerations and accelerations in rate. With a vertical error of ±0.9 m, the relatively wide "bandwidth" of the curve for the Delaware coast is clearly too coarse to discriminate the small amplitude (≤1 m) changes that probably have characterized the last several millennia. The classic oscillating sea level curve of Colquhoun stands in sharp contrast, belying the portrayal of a smooth (if gradually slowing) rise in late Holocene sea level, but, it is also illustrative of all the problems inherent in oscillating curves.

2.4.2   Oscillating Sea Level Curves: The South Carolina Coast

No sea level curve for the U.S. Atlantic Coast departs more from the present model for attempting to reconstruct a record of late Holocene sea level change than that of Colquhoun and his colleagues (Fig. 2.9). This curve, the first version of which was published almost 20 years ago (Colquhoun *et al.* 1980),

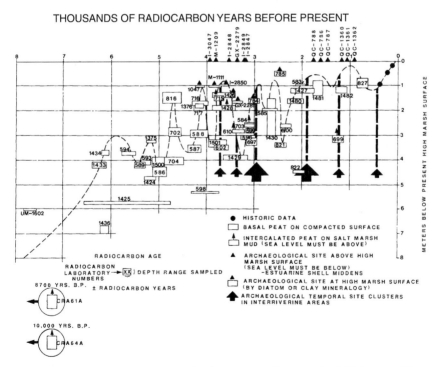

Figure 2.9   Holocene sea level curve for the South Carolina coast by Colquhoun and Brooks (1986). Notice, as in the Fairbridge curve for Brazil, that the archaeological evidence figures prominently in the reconstruction of sea level fluctuations, particularly former high stands.

incorporates data from historical sources, archaeological evidence, and basal and other marsh peats. Eight major transgressive phases, the earliest beginning about 6000 BP, are indicated in the curve and depict temporary high stands in mean sea level of almost 2 m. Significantly, however, none of these high stands were above modern sea level, though the most recent, dated around 1000 BP, appears to have been within a few decimeters of present limits.

Colquhoun *et al.* (1995) readily admit that some of the short transgressive phases may not have occurred (especially where there is little regional corroboration elsewhere) because of bad data. Perhaps less tractable than simply poor data is the acknowledged influence of hydrological variations in interriverine and upper reaches of the tributary estuaries where much of the supporting archaeological data (i.e., middens) appear to have been gathered. Though Colquhoun *et al.* (1981) presented persuasive arguments for the eustatic interpretation of spatial changes in the distribution of these sites (both midden and local resource exploitation sites), an alternate explanation could be that increased flooding during wetter climatic periods encouraged the adoption of sites somewhat higher above mean water level than usual. Exceptionally high water levels from tropical storms in the riverine and oligohaline reaches of major Coastal Plain estuaries can remain elevated for days (sometimes weeks; e.g., Tropical Storm Agnes in 1972 and Hurricane Floyd in North Carolina, September, 1999).

The incorporation of archaeological data, then, is the strength and weakness of the Colquhoun curve. Without these data (i.e., relying solely on the basal peats), the curve becomes another example of a smooth, nonoscillating curve (Fig. 2.9). It highlights the dilemma for attempting to go beyond basal peat curves: the evidence can be intriguing, but not necessarily defensible.

2.5   THE QUESTION OF HIGHER SEA LEVELS THAN PRESENT DURING THE LATE HOLOCENE

Perhaps no aspect of the famous "Fairbridge curve" (Fairbridge, 1961, 1976) so captured the attention of sea level researchers worldwide than the portrayal of sea levels higher than present throughout most of the late Holocene (see Fig. 2.1). This curve not only stood in sharp contradiction of a smooth gradual rise in sea level depicted by curves like that for the Delaware coast discussed above, but went further: modern sea level had been attained by an overall decline in sea levels over the last 5000 years. In fact, sea levels during the middle Holocene warm period were as much as 2 m higher than present.

Theoretical work on the influences of hydro-isostasy and glacio-isostasy (e.g., Clark *et al.,* 1978; see also Chapter 4) has indicated that the general trajectory of sea level changes shown in the Fairbridge curve is at least possible, if generally restricted to emerged beach ridges at higher latitudes around the Atlantic Ocean basin. On the other hand, for middle to lower latitudes above

30° N, theoretical considerations support a rising sea level that approached within a meter of modern limits by 5000 BP, and therefore slowly (if perhaps unevenly) rose to present mean sea level.

The work summarized above has important implications for any discussion of Holocene sea level highstands. First, it points to the primacy of geophysical causes for an apparent sea level high during the middle Holocene in many areas (especially around the North Atlantic). Second, by showing that sea levels were within "striking distance" of modern sea level relatively early, it clearly leaves open the possibility that sea levels could have exceeded present limits at some time during the late Holocene. After all, the rise would only have been on the order of less than a meter to reach present sea level and exceed it.

Reports of late Holocene sea level highstands recur in the sea level literature. Undoubtedly, early descriptions of such phenomena were marred by poor dating and inconclusive proof that the features in question (often beach ridges) were created by higher sea levels and not other factors like storms (see earlier discussion). But more recent reports are less likely to have these problems—the wide appreciation of the methodological traps among potential reviewers almost ensures that this would not be the case. Thus, such phenomena, like the ca. 2000 BP highstand discussed by Walker *et al.* (1995), have to be taken as real, at least in the sense that they exist. However, what they mean is another matter, particularly in terms of a eustatic response to climatic changes. Here there are considerable problems if one compares the level of climatic forcing that is necessary to achieve such rises by steric expansion of the oceans to the energy that was actually available.

Consider, as a reasonable approximation, that a 20-cm rise in global mean sea level requires at least 1°C increase in world mean temperature. This approximation is a whole ocean estimate and does not decouple surface waters of the euphotic zone from deeper ocean waters—which clearly is desirable, but brings into play such unknowns as rates of thermal diffusion with depth and other complexities. Nor does it consider shallow seas like the Caribbean, where light can reach the sea floor.

Reconstructions of climatic changes prior to the advent of instrumental records rely on proxy information that can have a high degree of imprecision. Nonetheless, there is sufficient agreement that temperatures during the warmest phases of the last several millennia were no more than 1°C higher than now. In fact, temperatures in east–central North America during the Hypsithermal ca. 8–5 ka BP, the warmest period of the Holocene, were just 2°C above present (Davis, 1984). Because this estimate is largely based on paleobotanical data (which generally reflect summer temperatures) and oceans warm up less than interior continental areas in summer, the degree of potential thermal expansion of the oceans even at this time was limited. Even recent computer models for a global sea level rise from a 2 to 2.5°C anthropogenic greenhouse warming fall considerably short of 1 m (Houghton *et al.*, 1996).

What, then, to make of reported late Holocene sea level highstands in excess of 1 m? First, the strict steric change argument is probably not supportable–the extensive work on late Holocene climate has yet to reveal a temperature increase during warm periods anywhere commensurate with this magnitude of sea level rise. In particular, those highstands within the last millennia purported to be on the order of 2–3 m above the mean sea level trend of any period certainly need reexamination; the possibility that the data are unreliable is simply too great. Second, alternative explanations for moderate highstands should be considered. Among these are geoidal variations, as first proposed by Morner (1976); local neotectonics (i.e., faults); and possibly tsunamis, from unsuspected offshore tectonism. Finally, the expectation that such highstands be widely correlated is perhaps unwarranted as it would be surprising if they were given the confounding effects of crustal dynamics and geoidal variations.

2.6   CONCLUSIONS

The controversy of an oscillating versus smooth rise in late Holocene sea level emerged almost immediately following publication of the first modern sea level curves in the 1950s. For most of the decades following, advances in both data collection and understanding of the climatological and geophysical frameworks for late Holocene sea level change did nothing to diminish the arguments. If anything, the increasing detail on late Holocene climates almost enjoined sea level researchers to find comparable variations in past sea level. And, in fact, sea level records at very fine temporal scales, generally supporting a fluctuating sea level, have appeared for several areas over the past 2 decades. For the past millennium, such records make clear the widespread flattening of the global secular sea level trend as the Little Ice Age reached its maximum in the 17th and 18th centuries (cf. Varekamp *et al.,* 1992).

Nevertheless, obtaining that equally detailed pictures of earlier sea level variations older than the last several millennia is uncertain. In many areas, coastal marshes, the best (i.e., least equivocal) archives of former sea levels, are seldom older than a few thousand years. It is rare to find a continuous marsh sediment record extending back 4000 to 5000 years, and then generally in the upper reaches of estuaries, where freshwater conditions (i.e., not clearly marine and tied to sea level) may have dominated the early part of the record. True salt marshes of the open coast (especially those in barrier lagoons) can be short-lived, tending to fall victim to some of the same forces produced by the transgression that created them.

Ultimately, length of record in intertidal deposits may be less of an issue than whether present relationships between accretion, sea level, and biological indicators have always been the same. This is a point raised by Pirazolli (1991), and it strikes at the heart of any expectations of increasing the resolution of sea level changes during the late Holocene prior to the past 2 millennia. Basal peat curves, though well tied to former sea level positions, are, as noted, essentially limited to portraying transgressions, not regressions. Geochemical and other indicators, as employed by Varekamp *et al.* (1992), can accommodate both rises and falls in sea level—the latter, where the sediments have not been destroyed by supertidal erosion or oxidation—but can give only the direction and relative magnitude of sea level change. Thus, improvements in fine scale temporal resolution of the late Holocene sea level history may rest in our understanding of how variations in sedimentary environment influenced sea level rise affects the ability of organisms to record those changes. As Scott and Medioli (1978) demonstrated over 20 years ago in the use of marsh foraminifera, the potential for detailed information on past sea level changes from biological indicators has yet to be fully exploited.

REFERENCES

Allen, E. A. (1977). Petrology and stratigraphy of Holocene coastal deposits along the western shore of Delaware Bay. Delaware Sea Grant Tech. Rept. DEL-SG-20-77, College of Marine Studies, University of Delaware, Newark.

Allen, J. R. L. (1987). Toward a quantitative chemostratigraphic model for sediments of late Flandrian age in the Severn Estuary. *Sedimentary Geol.* **53**, 73–100.

Allen, J. R. L., and Fulford, M. G. (1988). Romano-British settlement and industry on the wetlands of the Severn Estuary. *Antiquaries J.* **62**, 237–289.

Allen, J. R. L., and Rae, R. E. (1987). Late Fladrian shoreline oscillations in the Severn Estuary: a georphological and stratigraphical reconnaisance. *Phil. Trans. Roy. Soc.* **B315**, 185–230.

Belknap, D. F., and Kraft, J. C. (1977). Holocene relative sea-level changes and coastal stratigraphic units on the northwest flank of the Baltimore Canyon trough geosyncline. *J. Sedim. Petrol.* **47**, 610–629.

Belknap, D. F., and Kraft, J. C. (1981). Preservation potential of transgressive coastal lithosomes on the U. S. Atlantic shelf. *Marine Geol.* **42**, 429–442.

Bloom, A. L., and Stuiver, M. (1963). Submergence of the Connecticut coast. *Science* **139**, 332–334.

Chraztowski, M. J. (1986). Stratigraphy and geologic history of a Holocene lagoon, Rehoboth Bay and Indian River Bay, Delaware. Unpubl. Ph.D. dissertation University of Delaware, Newark, Delaware, 337p.

Clark, J. A., Farrell, W. E., and Peltier, W. R. (1978). Global changes in post-glacial sea level. *Quaternary Res.* **9**, 265–287.

Colquhoun, D. J., and Brooks, M. J. (1986). New evidence from the southeastern U.S. for eustatic components in the late Holocene sea levels. *Geoarchaeology* **1**, 275–291.

Colquhoun, D. J., Brooks, M. J., Abbott, W. H., Stapor, F. W., Newman, W. S., and Pardi, R. R. (1980). Principles and problems in establishing a Holocene sea level curve in South Carolina. In Howard, J. D. DePratter, C. B. and Frey, R. W. (eds.), *Excursions in Southeastern Geology. The Archaeology-Geology of the Georgia Coast. Geol. Soc. Am. Guidebook* **20**, Atlanta, 143–159.

Colquhoun, D. J., Brooks, M. J., Michie, J., Abbott, W. H., Stapor, F. W., Newman, W. S., and Pardi, R. R. (1981). Location of archaeological sites with respect to sea level in the southeastern United States. In Konigsson, L. K. and Paabo, K. (eds.), *Florilegium Florinis Dedicatum. Striae* **14**, 144–150.

Colquhoun, D. J., Brooks, M. J., and Stone, P. A. (1995). Sea-level fluctuation: Emphasis on temporal correlations with records from areas with strong hydrologic influences in the southeastern United States. In Finkl, C. W. Jr. (ed.), *Holocene Cycles*. Special issue, *J. Coastal Res* 191–196.

Cronin, T. M. (1987). Quaternary sea-level changes studies in the eastern United States of America: A methodological perspective. In Tooley, M. J. and Shennan, I. (eds.), *Sea Level Changes*. Basil Blackwell, London, 225–248.

Davis, M. B. (1984). Holocene vegetational history of the eastern United States. In Wright, H. E. Jr. (ed.), *Late-Quaternary Environments of the United States,* Vol. 2. *The Holocene*. Longman, London, 166–181.

Daly, R. A. (1934). *The Changing World of the Ice Age*. Yale Univ. Press, New Haven.

DePratter, C. B., and Howard, J. D. (1977). History of shoreline changes determined by archaeological dating; Georgia coast, United States of America. *Transactions of the Gulfcoast Association of Geological Societies* **27,** 251–258.

Donoghue, J. F., and Tanner, W. F. (1992). Quaternary terraces and shorelines of the Panhandle Florida region. In Fletcher, C. H. III and Wehmiller, J. F. (eds.), *Quaternary Coasts of the United States: Marine and Lacustrine Systems. Society for Sedimentary Geology Spec. Pub.* No. 48, Tulsa, Oklahoma. 233–242.

Easton, W. H. (1960). *Inveterbrate Paleontology*. Harper & Row, New York.

Emery, K. O. (1967). The Atlantic continental margin of the United States during the past 70 million years. *Geog. Assoc. Canada Spec. Paper* **4,** 53–70.

Fairbridge, R. W. (1961). Eustatic changes in sea level. In Ahrens, L. H. Press, F. Raukawa, K. and Runcorn, S. K. (eds.), *Physics and Chemistry of the Earth,* vol. 4. Pergamon, New York. 99–185.

Fairbridge, R. W. (1976). Shellfish-eating perceramic Indians in coastal Brazil. *Science* **191,** 353–359.

Fairbridge, R. W. (1987). The spectra of sealevel in a Holocene time frame. In Rampino, M. R. Sanders, J. E. Newman, W. S. and Kongsson, L. K. (eds.), *Climate: History, Periodicity and Predictability,* Van Nostrand, New York. 127–142.

Fairbridge, R. W. (1995). Some personal reminiscences on of the idea of cycles, especially in the Holocene. In Finkl, C. W. Jr. (ed.), *Holocene Cycles, J. Coastal Res. Special Issue No. 17,* 11–20.

Finkl, C. W., Jr. (1995). Holocene Cycles. *J. Coastal Res. Special Issue No. 17.*

Fischer, A. G. (1961). Stratigraphic record of transgressing seas in light of sedimentation on the Atlantic coast of New Jersey. *Bull. Am. Assn. Petrol. Geologists* **54,** 1656–1666.

Flint, R. F. (1971). *Glacial and Quaternary Geology*. John Wiley and Sons, New York.

Gribbin, J., and Lamb, H. H. (1978). Climatic change in historical times. In Gribbin J. (ed.), *Climatic Change*. Cambridge University Press, London. 69–82.

Grove, J. M. (1988). *The Little Ice Age*. Methuen, London.

Horner, R. W. (1972). Current proposals for the Thames and the organization of investigations. *Phil. Trans. Roy. Soc.* **A272,** 179–185.

Houghton, J. T., Meira Filho, L. G., Callander, B. A., Harris, N., Kattenberg, A., and Maskell, K. (eds.) (1996). *Climate Change 1995,* Intergovernmental Panel on Climate Change, Cambridge University Press, Cambridge.

Johnson, D. W. (1925). *The New England Shoreline*. Wiley, New York.

Kaye, C. A., and Barghoorn, E. S. (1964). Late Quaternary sea-level change and crustal rise at Boston, Massachusetts, with notes on the autocompaction of peat. *Geol. Soc. Am. Bull.* **75,** 63–80.

Kearney, M. S., and Stevenson, J. C. (1991). Island land loss and marsh vertical evidence for historical sea-level changes in Chesapeake Bay. *J. Coastal Res.* **7,** and 430–415.

Kennedy, J. (1999). Digging beneath Honolulu's Chinatown. *Natural History* **108,** 64–67.

Kidson, C. (1982). Sea level changes in the Holocene. *Quaternary Sci. Revs.* **162,** 121–151.

Kidson, C., and Heyworth, A. (1979). Sea "level". *Proc. 1978 Int. Symp. Coastal Evoln. in the Quaternary*. Sao Paulo, Brazil, 1–27.

Kraft, J. C. (1971). Sedimentary facies patterns and geological history of a Holocene marine transgression. *Geol. Soc. Am. Bull.* **82,** 2131–2158.

Kraft, J. C., Biggs, R., and Halsey, S. D. (1973). Morphology and vertical sedimentary sequence models in Holocene transgressive barrier systems. In Coates, D. R. (ed.), *Coastal Geomorphology*. State University of New York, Binghampton.

Kraft, J. C. (1976). Radiocarbon dates in the Delaware coastal zone (Eastern Atlantic Coast of North America): Delaware Sea Grant Technical Report (DEL-SG-19__76). College of Marine Studies, University of Delaware, Newark, Delaware.

Kraft, J. C., and Chrzastowski, M. J. (1985). Coastal Stratigraphic Seequences. In Davis, R. A. Jr. (ed.), *Coastal Sedimentary Environments*. Springer-Verlag, New York, 625–663.

Kraft, J. C., Chrzastowski, M. J., Belknap, D. F., Toscano, M. A., and Fletcher, C. H. (1987). Morphostratigraphy, sedimentary sequences and response to a relative rise in sea level along the Delaware coast. In Nummedal, D., Pilkey, O. H. and Howard, J. D. (eds.), *Sea Level Fluctuation and Coastal Evolution*. Tulsa, Oklahoma: Society of Economic Paleontologists and Mineralogists *Spec. Pub.* No. 41, 129–144.

Long, D., Dawson, A. G., and Smith, D. E. (1989). A Holocene tsunami deposit in eastern Scotland. *J. Quaternary Sci.* **4,** 61–66.

Morner, N.-A. (1973). Eustatic changes during the last 300 years. *Palaeogeography, Palaeoclimatology, Paleoecology* **13,** 1–14.

Morner, N.-A. (1976). Eustasy and geoid changes. *J. Geol.* **84,** 123–152.

Mudge, B. F. (1858). The salt marshes of Lynn. *Proc. Essex Inst.* **2,** 117–119.

National Academy of Sciences. (1987). Responding to Changes in Sea Level. National Academy of Sciences Press, Washington, DC.

Newman, W. S., Cinquemani, L. J., Pardi, Richard R., and Marcus, L. F. (1980). Holocene delevelling of the United States' East Coast. In Morner, N.-A. (ed.), *Earth Rheology, Isostasy and Eustasy*. John Wiley, New York, 449–463.

Oertel, G. F., Kraft, J. C., Kearney, M. S., and Woo, H. J. (1992). A rational theory of barrier-lagoon development. In Fletcher, C. H. III and Wehmiller, J. F. (eds.), *Quaternary Coasts of the United States: Marine and Lacustrine Systems. Soc. Sedimentary Geol. Spec. Pub. No. 48,* 77–88.

Plassche, O. van der. (1977). A manual for sample collection and evaluation of sea level data. Free University, Amsterdam.

Pirazolli, P. A. (1991). *World Atlas of Holocene Sea-Level Changes*. Elsevier, Amsterdam.

Rampino, M. R., and Sanders, J. E. (1981). Episodic growth of Holocene tidal marshes in the northeastern United States: A possible indicator of eustatic sea-level fluctuations. *Geology* **9,** 63–67.

Redfield, A. C. (1972). Development of a New England salt marsh. *Ecological Monographs* **42,** 201–237.

Reineck, H.-E., and Singh, I. B. (1975). *Depositional Sedimentary Environments*. Springer-Verlag, New York.

Riggs, S. R., York, L. L., Wehmiller, J. F., and Snyder, S. W. (1992). Depositional patterns resulting from high-frequency Quaternary sea-level fluctuations. In Fletcher, C. H. III and Wehmiller, J. F. (eds.), *Quaternary Coasts of the United States: Marine and Lacustrine Systems. Soc. Sedimentary Geol. Spec. Pub.* No. 48, 141–154.

Scott, D. B., and Medioli, F. S. (1978). Vertical zonations of marsh foraminifera as accurate indicators of former sea levels. *Nature* **272,** 528–531.

Shaler, N. S. (1885). Preliminary report on seacoast swamps on the eastern United States. *U.S. Geol. Survey 6th Annl. Rept.* 359–398.

Shennan, I. (1987). Holocene sea-level changes in the North Sea region. In Tooley, M. J. and Shennan I. (eds.), *Sea-Level Changes*. Basil Blackwell, Oxford, 109–151.

Silberhorn, G. M. (1982). Common plants of the mid-Atlantic Coast: A field guide. Johns Hopkins Univ. Press, Baltimore.

Stapor, F. W., Jr. (1973). Coastal sand budgets and Holocene beach ridge plain development, northwest Florida. Ph.D. thesis, Florida State University, Tallahassee.

Stapor, F. W., Jr., Mathews, T. D., and Tanner, W. F. (1991). Barrier-island progradation and Holocene sea-level history in southwest Florida. *J. Coastal Res.* **7,** 815–838.

Stevenson, J. C., Ward, L. G., and Kearney, M. S. (1986). Vertical accretion rates in marshes with varying rates of sea-level rise. In Wolfe, D. A. (ed.), *Estuarine Variability.* Academic Press, New York, 241–259.

Stumpf, R. P. (1983). The process of sedimentation on the surface of a salt marsh. *Estuaries, Coastal and Shelf Science* **17,** 495–508.

Tooley, M. J., and Shennan, I. (eds.). (1987). *Sea-Level Changes.* Basil Blackwell, Oxford.

Varekamp, J. C., Thomas, E., and Plassche, O. van de (1992). Relative sea level rise and climate change over the last 1500 years. *Terra Nova* **4,** 293–304.

Walker, K. J., Stapor, F. W., Jr., and Marquardt, W. H. (1995). Archaeological evidence for a 1750–1450 BP higher-than-present sea level along Florida's Gulf Coast. In Finkl, C. W., Jr. (ed.), *Holocene Cycles. J. Coastal Res. Special Issue No. 17,* 205–218.

Ward, L. G., Kearney, M. S., and Stevenson, J. C. (1998). Variations in sedimentary environments and accretionary patterns in estuarine marshes undergoing rapid submergence, Chesapeake Bay. *Marine Geology* **151,** 111–134.

Whittle, A. W. R. (1989). Two Bronze Age occupations and an Iron Age channel on the Gwent foreshore. *Bull. Celtic Board of Studies* **36,** 200–223.

Woodworth, P. L. (1999). A study of changes in high water levels and tides at Liverpool during the last two hundred and thirty years with some historical background. Proudman Oceanographic Report No. 56, Proudman Oceanographic Laboratories, Bidston, U.K.

Chapter 3 | Sea Level Change in the Era of the Recording Tide Gauge

Bruce C. Douglas

3.1 INTRODUCTION

Sea level rise is often regarded as an issue for the future, rather than of the present day. The most alarming scenarios involve partial melting or collapse of a great polar ice sheet a hundred years or more hence due to global warming. The result of such a catastrophe would be flooding of coastal regions of the world, causing massive loss of life and property and breakdown of social order. Fortunately, plausible forecasts (Houghton *et al.,* 1996) of sea level change for the 21st century are far less ominous. But as Chapters 2 and 8 document, the impact of even a moderate rate of sea level rise is severe, especially for island and developing nations. In fact, if global sea level rises in the 21st century at only the 20th-century rate of about 2 mm per year, the economic and social burdens will still be profound, because an increase of sea level significantly increases the impact of storms on heavily populated low-lying coastal areas. It is a matter of practical urgency to determine the amount and causes of global sea level rise so that mitigation activities can begin as soon as possible.

The global increase of sea level by as much as 20 cm over the last century, although important, is minor compared to what occurred in the distant past. Sea level rose by approximately 125 meters (over 600 times greater!) as a result of the disappearance of the great glacial ice sheets that began about 21,000 years ago. Melting was complete everywhere by 4000–5000 years ago, at which time relative sea level was within a few meters of its present value for most of the earth. Removal of the ice load also caused the elevation of certain high-latitude locations to increase by hundreds of meters and adjacent areas to subside as the earth adjusted viscoelastically. This change of elevation, referred to as glacial isostatic adjustment (GIA), is a global phenomenon that continues to this day everywhere at rates that are significant in some areas in comparison to estimates of 20th-century global sea level rise.

The last few thousand years are especially interesting as far as global sea level rise is concerned. Geological and other evidence presented in Chapter

2 and by Gornitz (1995a) and Varekamp *et al.* (1992) suggest that during this recent period, the *average* rate of change of global sea level has been very small, much less than the 20th-century rate. This conclusion has also been reached by Flemming (1978, 1982) using a novel method. He analyzed elevations of hundreds of Mediterranean coastal archeological sites relative to modern sea level and concluded that the average rate of sea level rise during the last 2 millennia has been only 0.2 mm/yr. Apparently the 20th-century rate of about 2 mm per year is an historically recent development; one could even be tempted to attribute the increased rate to the 20th-century global warming of approximately 0.3–0.6°C (Houghton *et al.,* 1996). But sea level began to rise at the much faster modern rate near the mid-19th century, too soon for 20th-century warming to be held responsible.

This is not to say that global warming (or cooling) has a negligible effect on sea level. A crude estimate of its importance can be calculated from the thermal expansion coefficient of sea water. This quantity varies strongly with salinity and temperature (water is a peculiar substance—consider what happens when it freezes), but for tropical and midlatitude ocean temperatures the value differs from $2.5 \times 10^{-4}/°C$ by less than $\pm 50\%$. Using this value of the expansion coefficient gives an increase in the level of a 1000-m layer of ocean of the order of 10 cm for a temperature change of 0.5°C. So thermal expansion as a source of sea level change is not to be ignored. Detailed calculations (summarized by Warrick *et al.,* 1996) based on the actual structure of the oceans indicate that the contribution of thermal expansion to global sea level in the 20th century is on the order of 2–7 cm.

Another potential source of increased sea level from global warming often mentioned in the media is melting of the polar glacial ice caps. But the great ice sheets of Greenland and Antarctica are unlikely to have been affected by a change of a few tenths of a degree Celsius in a century. Ice is an excellent insulator, and the thermal inertia of the polar caps is very great. However, smaller Alpine-type glaciers are likely affected. Meier (1984, 1990) estimated that mountain ice sheets are retreating enough to cause global sea level to rise 4.6 ± 2.6 cm in the 20th century.

Other questions abound that relate to global sea level rise and global warming. For example, the Greenland and Antarctic ice sheets can play a role other than melting. In the case of Greenland, Zwally *et al.* (1989) claimed that satellite altimeter data showed a thickening of ice there during 1978–1987 equivalent in volume to a fall of sea level of a few tenths of a millimeter per year. If this scenario held true over an extended time, a possible increase of global sea level from thermal expansion or other sources could be offset to some extent by increased storage of water in the form of ice. However, a later reanalysis of the satellite altimeter data with refined methods by Davis *et al.* (1998) showed a negligible change of ice elevation for Greenland. But the issue remains. If global warming causes increased precipitation at high latitudes with concomitant storage of water in the form of ice, sea level rise due to thermal expansion of the ocean or melting of small glaciers could be offset to a certain extent. The ice mass balance (i.e., whether or not there is net accretion or loss of ice) of Greenland and Antarctica is of critical importance for accounting for changes of global sea level. However, Warrick *et al.* (1996) concluded that because of uncertainty of the ice mass balance of the Greenland and Antarctic ice sheets, not even the sign of their contribution to 20th-century sea level rise can be determined. But there is still concern that a response is possible.

An issue unrelated to global warming that affects the rate of global sea level rise is storage and mining of water. This question is dealt with in detail in Chapter 5. In summary, since World War II the increase in the amount of water held in large and small reservoirs (and kept out of the oceans), in combination with other effects on the hydrological cycle, has a sea level rise equivalent estimated to be 0.9 ± 0.5 mm/yr. This is enough to offset an increase in the rate of sea level rise that could have occurred in the last 50 years due to global warming.

Warrick *et al.* (1996) summarized the factors affecting sea level rise in the 20th century. Under the assumption that the mass balance of the Greenland and Antarctic ice caps is zero, they concluded that the remaining contributors to global sea level could add up to 0.8 mm/yr of sea level rise in the 20th century, with a substantial uncertainty. This value is much smaller than most estimates of global sea level rise based on tide gauge data. Table 3.1 shows the values published in the last decade. They range in magnitude from 1.0 to 2.4 mm/yr. All of these authors used tide gauge data from the same data base, so the differing estimates reflect selection

Table 3.1

Recent Determinations of Global Sea Level Rise from Tide Gauge Data

Author	Estimate (mm/yr)	Comments
Peltier and Tushingham (1989, 1991)	2.4 ± 0.9^{a}	Global data
Barnett (1990)	1–2	Global data
Nakiboglu and Lambeck (1990)	1.15 ± 0.38	Global data
Trupin and Wahr (1990)	1.75 ± 0.13	Global data
Douglas (1991)	1.8 ± 0.1	Global data
Shennan and Woodworth (1992)	1.0 ± 0.15	U.K. and Europe
Mitrovica and Davis (1995)	1.1–1.6	Global data
Davis and Mitrovica (1996)	1.5 ± 0.3	U.S. east coast
Peltier (1996)	1.94 ± 0.6^{a}	U.S. east coast
Peltier and Jiang (1997)	1.8 ± 0.6^{a}	U.S. east coast
Douglas (1997)	1.8 ± 0.1	Global data

[a] Standard deviation of trends about their mean. The formal SE is a few tenths.

criteria and data analysis methods, particularly record length and the correction for glacial isostatic adjustment (GIA). Warrick *et al.* (1996) selected 1.8 mm/yr as the "best estimate," with the scatter of all of the estimates (1–2.4 mm/yr) taken to reflect in some measure the uncertainty of the rate. However, Douglas (1995, 1997) and Peltier and Jiang (1997) (see also Chapter 4 of this volume) conclude that there is a preponderance of evidence that the 20th-century rate of sea level rise is much closer to 2 than to 1 mm/yr. To resolve these issues of the contributors to sea level change and its observed rate of rise, at least two kinds of continuing research are needed. First, estimates of the contributions to sea level rise by thermal expansion of the ocean, melting of small glaciers, impoundment of water in large and small reservoirs, mining of groundwater, and accumulation or loss of ice on the great polar ice sheets must be improved. Second, the rate of global sea level rise for the 20th century determined from tide gauge measurements can be further refined and monitored by altimeter satellites to determine if changes in the rate are occurring. Any change in the rate of sea level rise is potentially detectable much sooner by altimetric satellites than tide gauges because of the former's near-global coverage of the oceans. Satellite measurements of sea level variation are the subject of Chapter 6. It is the purpose of the present chapter to examine the problems encountered in determining an accurate estimate of the rate of global sea level rise from tide gauge data.

3.2   TIDE GAUGES AND THEIR DATA

A tide gauge in its most basic form is a simple graduated staff or other reference on which the water level can be estimated visually. Such readings are available for a few sites as far back as the 17th century. Measurement series made this way can be surprisingly accurate if carefully and faithfully done, and tide staffs are used to this day to provide an independent check on the functioning of many tide gauges. However, to be useful for most investigations of climate change, a record of water levels should be temporally dense and long (the longer the better), and without the limitations of visual estimates. In addition, the record must maintain internal consistency in spite of repairs, replacements, and changes of tide gauge technology. With these requirements in mind, what can be considered the modern epoch of sea level measurement began about 165 years ago with the use of a recording tide gauge at Sheerness in the United Kingdom. Other, more continuous series of sea level observations began soon after in the Baltic region and elsewhere in northern Europe. The longest continuous U.S. sea level record from a recording tide gauge began in San Francisco in 1854. However, there are not very many really long tide gauge records;

consider Fig. 3.1. This histogram shows the number of tide gauge records of various lengths.

It is clear from the cumulative percentages shown on the right-hand side of Fig. 3.1 that the vast majority (80%+) of records are less than 60 years long and that only a relative handful cover the entire 20th century. In addition, the longest records are nearly all from sites in northern Europe and the United States, so the world's oceans are not well sampled. The lack of very many useful and very long records will be seen below to be one of the most important issues in determining a rate of global sea level rise for the 20th century.

In addition to any global change of sea level, a tide gauge records (of course) tides, effects of ocean circulation, meteorological forcing of the water, local or regional uplift or subsidence at the measurement site, vertical movements of the platform to which the gauge is mounted, and errors inherent to the gauges. With so many factors involved in maintaining a consistent record over an extended time such as a century or more, a cooperative international program is a necessity. The world's focal point for sea level data and information is the Permanent Service for Mean Sea Level (PSMSL), based at the Proudman Oceanographic Laboratory in the United Kingdom. The PSMSL manages sea level data from nearly 200 national authorities. It also sets standards, provides training in operation and maintenance of instrumentation, and carries out important sea level research. Their Web site found at http://www.pol.ac.uk provides a wealth of information about sea level data, measurement technology, data analysis, and links to numerous important *in situ* and artificial satellite observing programs.

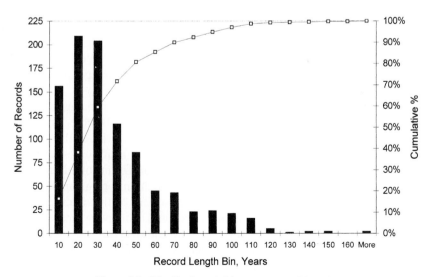

Figure 3.1   Distribution of tide gauge record lengths.

Tide gauges have taken many forms. As noted, the simplest measurement technique is to observe visually the level of the water on a graduated staff. Such readings are individually accurate to only a few centimeters, but if done frequently can give estimates of daily tidal range adequate for marine navigation and good estimates of annual mean water level. Much better results suited for scientific use are obtained by measuring water level in a so-called *stilling well*. A common form of this device used in the United States consists of a vertically standing 30-cm-diameter pipe that cones down to a 2.5-cm-inlet orifice. The small-diameter hole available for water entry and exit effectively limits the flow of water in and out of the well over time intervals (seconds) associated with waves, but does not interfere with longer period flow variations such as those from tides. The orifice thus serves as a mechanical low-pass filter that eliminates high-frequency variations of water level in the well. Water level in the stilling well has traditionally been determined by means of a float connected by a wire to a recording device. In more recent years, echo sounding of the distance from an audio or radar source to the water surface has been used successfully. Typically, the values of water level recorded are 6-minute averages of the actual measurements.

It is also possible to measure variations of water level from changes in the weight of the water overlying a special quartz crystal oscillator which responds in a manner proportional to the pressure change. Another means of measuring water level variations is a "bubbler" gauge. A flow of nitrogen or compressed air into a tube with one end in the water is adjusted continuously by a mechanism so that it releases bubbles at a constant rate. When the water level rises, the pressure must be increased to maintain the flow of bubbles. The pressure variations are thus a measure of water level. A small tank of gas can operate this system for many months. These systems are precise and simple and are often used to supplement other gauge types to provide short-term backup capability. For more and detailed information about tide gauge history and technology, consult the U.S. National Ocean Service Web site http://www.opsd.nos.noaa.gov/ and the PSMSL site at http://www.pol.ac.uk/psmsl/training.

In common with most scientific measurements, it is easier to measure short-term variations of water level than very long-period ones. Bubbler and quartz crystal systems can drift with time through transducer drift, and so are not suitable for investigation of long-term sea level rise at a site unless equipped with ancillary datum control equipment. Float-type and echo sounding systems have their own problems, but are well suited to monitoring long-term sea level trends if properly maintained. A critical aspect of this maintenance is ensuring that the gauge, usually mounted on a pier, is not giving measurements contaminated by settling or other movement of the structure. In addition, the gauge must be remounted or replaced in a known manner that enables continuity of the record to be maintained. To solve these problems, one or more geodetic benchmarks are installed near the pier on presumably stable ground.

Periodic (usually annual) geodetic surveys are made to a gauge to determine if any vertical changes in the gauge mount have occurred. Replacing or moving a gauge without introducing a "step" discontinuity in the record is also made possible using the geodetic reference benchmarks. The PSMSL refers to gauge records for which a complete geodetic history is available as Revised Local Reference (RLR) sites. The PSMSL states that with only a few exceptions, RLR sites are the ones that should be used for scientific applications.

3.3   INTERPRETING SEA LEVEL RECORDS

The seeming inability of investigators during the last few decades to arrive at a consensus concerning the precise rate of global sea level rise, or even how to approach the problem, has led some authors to conclude that global sea level rise cannot be measured at all. Barnett (1984) states that ". . . it is not possible to uniquely determine either a global rate of change of sea level or even the average rate of change associated with the existing inadequate data set." Emery and Aubrey (1991) state that (p. 176) "At present, we cannot discover a statistically reliable rate for eustatic rise of sea level alone . . . ." Pirazzoli (1993) is the most pessimistic, declaring ". . . the determination of a single sea-level curve of global applicability is an illusory task." Douglas (1995) and Gornitz (1995b) provided detailed analyses and interpretation of estimates published since 1980, and concluded that the situation is not so bleak. In his determinations of global sea level rise and (lack of) acceleration, Douglas (1991, 1992, 1997) stressed the importance of using very long records. He concluded that the large differences in estimates of sea level rise could be explained in nearly all cases by the selection criteria used by investigators, especially the use of short (<60 years) records and data from sites affected by plate tectonics. Gornitz (1995b) further pointed out that the use of long records and explicit model corrections for glacial isostatic adjustment (GIA) in most cases leads to values nearer to 2 than to 1 mm/yr. But controversy continues. In this section we examine carefully some individual tide gauge records that show explicitly how the characteristics of sea level change can radically influence the results for the computed sea level trend at a site. In particular, the role of record length will be shown to be of the utmost importance in determining trends of relative sea level.

   To measure water level appears to be simple, but a time series of sea levels measured by a tide gauge contains much information. As noted, the gauge is designed to "filter out" some information, that is, changes of water level occurring over periods of a few seconds due to waves; these changes are better measured by other instruments. Tide gauges respond well to the familiar daily and semi-daily tides, whose amplitude is typically of the order of a meter. This is 500 or more times the amount of the yearly rise of sea level at most places, so removal of tidal signals from sea level data by a low-pass filter is necessary for investigations of long-term sea level change. This can be done very precisely if desired, since the motions of the sun and moon and rotation of the earth determine tidal periods (Pugh, 1987). Simpler digital filtering techniques also suffice. Some other fluctuations of coastal sea level, such as shallow water effects of tides, wind-driven setup, storm surges, or precipitation runoff, are not so easily modeled. However, averaging the water level over monthly or annual intervals effectively eliminates these small or limited duration events. Other larger and temporally more enduring water level variations, such as the seasonal-interannual fluctuations due to El Niño, create significant difficulties in determining a trend of sea level.

As an example of the signal content of a sea level record, consider the monthly mean water levels in Fig. 3.2. These data from the tide gauge site at the Scripps Institution of Oceanography in La Jolla, California, were obtained from the PSMSL Web site. The PSMSL provides monthly mean data in columnar format, making it very easy to analyze the data with a spreadsheet program. The Scripps gauge, operated and maintained as a part of the National Water Level Network by the National Ocean Service (NOS) of the National Oceanic and Atmospheric Administration (NOAA), is an RLR site that has been in operation since 1925. It is located at the end of a long research pier facing the open ocean, and provides an excellent record of sea levels.

The data shown in Fig. 3.2 are *relative* sea levels (RSLs). This merely means that the origin is arbitrary and that changes in water level cannot be distinguished from vertical movements of the land or platform on which the

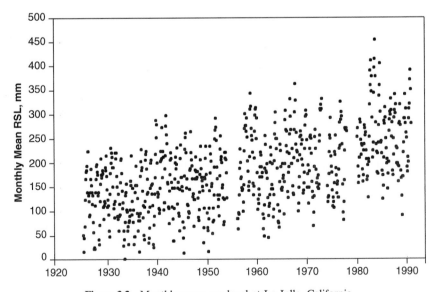

Figure 3.2 Monthly mean sea level at La Jolla, California.

gauge is located. An *absolute* system of sea levels, one in which all tide gauges are positioned in a common coordinate system, is highly desirable. If the coordinates of a tide gauge could be independently monitored for change in an absolute global geodetic reference frame, the observed trend of sea level at a location could be freed of contamination from vertical land movement. To this end, there is an international effort under way to provide such a system at the millimeter level of precision using a wide variety of modern geodetic technologies. For more information, consult the Harvard—Smithsonian Center for Astrophysics Sea Level Page at http://rgalp6.harvard.edu/index_rsl.html, the International GPS Service for Geodynamics Web site at http://igscb.jpl.nasa.gov/, and the International Earth Rotation Service at http://hpiers.obspm.fr/. As discussed in Chapter 6, the results are promising. Within a decade it appears that vertical crustal movements at enough tide gauge sites will be known accurately enough in an absolute geodetic reference frame to solve the vexing problem of local and regional elevation changes of the land.

The monthly mean sea levels in Fig. 3.2 display an upward trend of a few millimeters per year, with an additional "noise" of about 100 mm. The real nature of the high-frequency component of the record is obscured because of the scale of the graph. Figure 3.3 shows the data for only the decade 1980–1990, along with a 3-month running mean curve. What may have looked like random noise in Fig. 3.2 is revealed in Fig. 3.3 to be very systematic in character. The obvious *annual* component of the signal arises (at least at this particular location) largely from thermal expansion and contraction of the

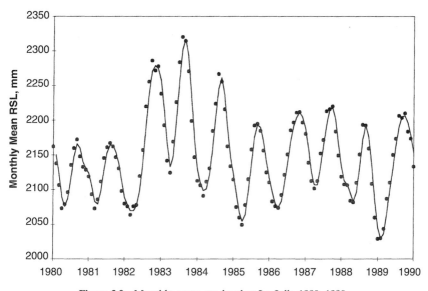

Figure 3.3 Monthly mean sea level at La Jolla 1980–1990.

upper layer of the ocean; the remaining components are small and irregular and probably meteorological in origin. Also easily visible as a superimposed multiyear anomalous period of elevated sea level is the effect of the 1982–1983 El Niño event. El Niño is primarily a warming phenomenon of the tropical Pacific Ocean, but also produces changes of sea level that propagate north and south along the west coast of North and South America (Chelton and Davis, 1982) raising sea level well beyond the norm. Higher water level during El Niño can have important consequences. Large waves generated by Pacific storms struck the California coast during times of unusually high sea level in the winter of 1982–1983 and caused severe beach and bluff erosion in some areas.

For an interesting example of decadal and longer variations of sea level, consider Fig. 3.4. It displays annual mean sea levels for Annapolis and Baltimore, Maryland, and Atlantic City, New Jersey, since 1930. The sea levels are in units of millimeters, and only *relative* changes can be inferred from each series. In addition, the series are offset by an arbitrary amount from one another for the sake of clarity and each has an added running 3-year mean through the data. The high correlation of the series at long periods is very obvious, and gives confidence that the multiyear fluctuations are real. (In Chapter 7 Sturges and Hong provide the explanation for them; their origin is meteorological.) Multiple sea level records from nearby tide gauges are useful for verifying that details of a record of sea levels are real, and not a result of instrument error.

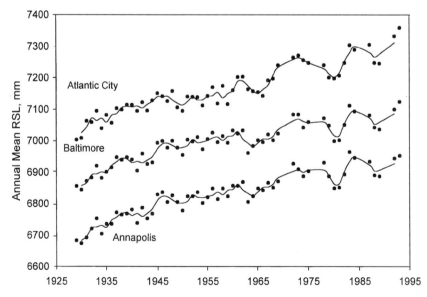

Figure 3.4   U.S. mid-Atlantic annual mean relative sea levels.

What may be surprising about the series in Fig. 3.4 is that an interannual correlation of sea levels actually exists in these records. Atlantic City is on the open coast, and Annapolis and Baltimore are located in different branches of a very large estuary (the Chesapeake Bay) that receives heavy seasonal inflows of fresh water from rivers. But viewed from the perspective of interannual and longer periods of water level variation, Chesapeake Bay water level is in dynamic equilibrium with the nearby open ocean. There are a number of other gauges in the Bay, and they all display this same correlation at interannual and longer periods. Douglas (1991, 1992, 1997) evaluated the tide gauge records in his determinations of global sea level rise and acceleration in this manner and placed them into morphological groups based on their apparent correlation at low frequencies with their neighbors. This ensured that no oceanographically coherent region received undue weight in the global average of regional sea level trends simply because it happened to have a large number of tide gauge sites.

Sometimes relatively nearby gauge site records disagree in their interannual variations. When this occurs, there is always the possibility that one of the records is not suitable for some reason. A good example of this problem is given by a pair of long Australian sea level records. In his investigation of global sea level rise, Douglas (1991) found that the long-period variations of relative sea levels at Sydney and Newcastle, only a few 100 km apart, were 180° out of phase for nearly half of their common data interval. The trends derived from these records also differed by a factor of two. These records were not used in his analysis, nor in his reanalysis (Douglas, 1997) because no explanation could be found for the difference.

Another pair of nearby tide gauges, Buenos Aires and Quequen in Argentina, also give sharply different trends (1.5 vs 0.8 mm/yr) over their rather long (85 and 65 years, respectively) records. In addition, the records of the two sites do not in any way resemble each other morphologically even though the gauges are only about 400 km apart. Fortunately, a plausible explanation for the difference can be found in this case. It is a surprising one and offers an excellent illustration of the problems involved in obtaining a reliable trend from a series of water levels recorded by a tide gauge.

Figure 3.5 displays annual values of mean sea level at Buenos Aires, with the final 7 years (1981–1987) shown as open squares; these data after 1980 were not used to determine the linear regression trend result of 1.1 mm/yr shown in Fig. 3.5. It is remarkable that eliminating only a small part of the record (i.e., the 7 years after 1980) decreased the estimated trend of water level from 1.5 to 1.1 mm/yr, much closer to the Quequen value of 0.8 mm/yr. What caused the very rapid rise of water level after 1980? Figure 3.6 provides the probable answer. This figure displays detrended and normalized values of smoothed monthly means of the Buenos Aires sea level record and the negative of the Southern Oscillation Index (SOI). The SOI, which is computed from the difference of barometric pressures at Tahiti and Darwin, is negative

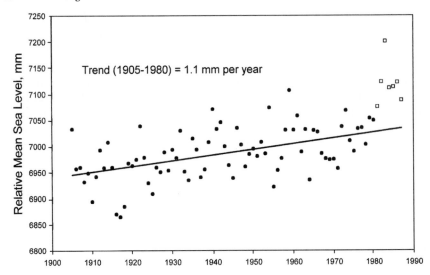

Figure 3.5   Annual mean relative sea level at Buenos Aires.

during warm El Niño–Southern Oscillation (ENSO) events and positive during cold ones. See Diaz and Markgraf (1992) and Trenberth (1997) for more detailed discussions of the ENSO phenomenon. The CD-ROM accompanying this book contains a sequence of global monthly mean sea surface temperature

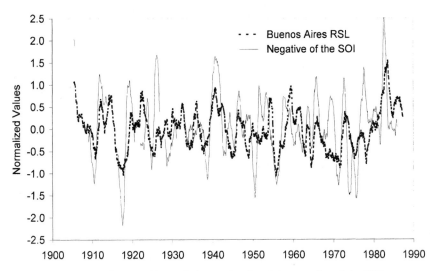

Figure 3.6   Buenos Aires relative sea level and the negative of the SOI.

anomaly maps from 1982 to 1999 prepared by Dr. Charles Sun of the National Oceanic and Atmospheric Administration (NOAA). These maps are shown in rapid succession and present a "movie" of the phenomenon that clearly illustrates the global extent of it. The most up-to-date version of the animation can be found at www.noaa.nodc.gov.

SOI values used to prepare Fig. 3.6 were obtained from the NOAA Climate Prediction Center (CPC) Web site http://www.cpc.ncep.noaa.gov. The signs have been reversed on the SOI values in Fig. 3.6 in order to make clearer the correlation of the SOI and sea level. There are some gaps in the SOI record, but Fig. 3.6 shows that there is an obvious correlation ($r = 0.46$) of sea level at Buenos Aires with the SOI. Some of the largest ENSO events are especially well reflected in sea level fluctuations there. Figure 3.7, the equivalent of Fig. 3.6 for Quequen relative sea levels, in contrast shows no significant correlation ($r = 0.03$) with the SOI. How is this disagreement between Buenos Aires and Quequen sea level records and the SOI possible? The answer almost certainly lies in the geographic situation of these two sites. Quequen is south of Buenos Aires on the open South Atlantic Ocean. No correlation with ENSO events is seen there, nor expected; for one to occur in analogy to the effect of the ENSO on western North and South American water levels, a coastal wave would have to propagate around the southern tip of South America and up the east coast, a physically unrealistic scenario. Buenos Aires has a different situational aspect. It is located well inside an estuary that drains a very large portion of northern Argentina, southern Brazil, Paraguay, and Uruguay. Increased rainfall during ENSO years would have the effect of decreasing the

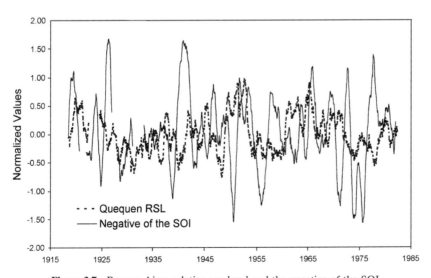

Figure 3.7   Buenos Aires relative sea level and the negative of the SOI.

salinity of the water in the estuary and increasing its flow, both leading to an increase of water level. So the correlation can plausibly be attributed to an indirect ENSO effect, rather than a direct one. But the impact on water levels at Buenos Aires is nearly as dramatic as seen on U.S. Pacific Coast water levels.

The example of Buenos Aires and Quequen demonstrates that the usual approach to determining a trend of sea level, ordinary linear regression, has significant limitations. Specifically, the linear regression model assumes that the signal or time series consists of a trend with added Gaussian (i.e., normally distributed) random noise. To the extent that a series of water levels deviates from this model, the computed trend and its uncertainty will be corrupted. Thus the linear regression model is at best an approximation to the real nature of a series of water levels.

From a statistical point of view, there are two main issues in using linear regression to determine sea level trends from tide gauge records. First, it is apparent from the examples in this chapter that there is serial correlation of the measurement residuals about the trend line. This demonstrates that the measurements contain information that cannot be accommodated by the model, so that the effective number of degrees of freedom is less than the number of observations minus two. Second, the character of the residuals is inconsistent. They seem to lack the character called *stationarity,* that is, their statistical properties at least appear to be nonuniform over the series. (Wunsch (1999) has pointed out very clearly how such appearances can be deceptive.) In these finite length water level records, the lack of uniformity of the signal character over the record creates great difficulties in identifying and determining an underlying trend.

To illustrate the problems of model inadequacy, it is instructive to analyze the Buenos Aires sea level trend in terms of formal uncertainties, and compare to actual results. The standard deviation, σ, of the trend obtained from a linear regression is calculated from (see any elementary statistics text)

$$\sigma^2_{trend} = \text{MSE}/[(\Sigma(t_i - t_{mean})^2], \tag{1}$$

where MSE is the sum of squares of the residuals about the trendline divided by the number of observations less 2, and the summation is over all values of the times t_i. Application of this formula for the standard deviation of the trend at Buenos Aires gives $\sigma = 0.22$ mm/yr for both 1905–1980 and 1905–1987. Fortuitously, the shorter span of data from 1905–1980 gives the same uncertainty (to two significant figures) of the trend as the longer span. This occurs because even though shorter, the MSE is smaller due to elimination of the large residuals after 1980. But the actual results for the trend upon extending the record from 75 years (1905–1980) to 83 years (1905–1987), only 9%, increases the apparent trend of sea level by about 36% (1.1 to 1.5 mm/yr). Obviously, the formal uncertainty of the trends obtained from Eq. (1) is too small. The explanation for this anomaly lies in the nature of the sea level signal. We have noted that the residuals about the trend line are in fact serially correlated; that is, adjacent residuals have a tendency to be similar in size. In addition, the anomalous increase of sea levels after 1980 is inconsistent with the assumption of stationarity, and has an extreme impact on the trend estimate.

A quantitative measure of the serial correlation of the residuals about the trend line for Buenos Aires sea levels shown in Fig. 3.5 is presented in Fig. 3.8, where members of adjacent residual pairs (i.e., lag 1) are plotted against each other. If there were no serial correlation, there would be no trend; the existence of the trend shows that there is a general tendency for adjacent residuals to increase or decrease together. For a discussion of serial correlation in residuals, see Draper and Smith (1981).

Maul and Martin (1993) used the lag 1 correlation of residuals to analyze their long series of annual mean sea levels at Key West, Florida, and found that the effective sample size was less than one-half the number of observations in that case. The results shown here for Buenos Aires are similar. The lag 1 serial correlation indicates an increase of the formal error by about 40%, that is, to about 0.3 mm/yr, which is significant, but not enough to explain the extreme sensitivity of results to inclusion or exclusion of 7 years of data after 1980. That sensitivity mostly exists because of the anomalous increase of sea level near the end of the record associated with the extreme 1982–1983 ENSO event.

In regard to stationarity of the residuals, the assumption of ordinary linear regression is violated by the unpredictable large anomalies caused by ENSO events (or anything else). The anomaly of sea levels at the end of the Buenos Aires record has an especially large effect on the apparent overall trend of sea level because of its location in the series. If an anomaly is near the average of the independent variable, there is little effect on the derived trend, but if

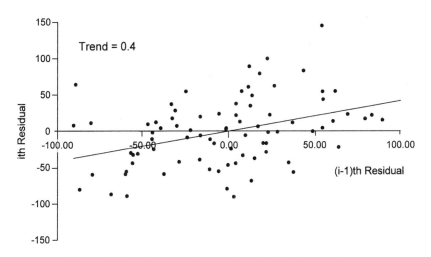

Figure 3.8 Lag 1 autocorrelation of residuals for Buenos Aires.

it is near either end of the series, the effect can be very great. Should the data after 1980 be discarded? In this case, we can identify the reason for the anomaly (i.e., a large ENSO event) and use that knowledge to justify elimination of the anomalous data. The more reasonable estimate of the underlying long-term trend of sea level at Buenos Aires is the one lacking data after 1980, that is 1.1 rather than 1.5 mm/yr.

The effect of El Niño on the Buenos Aires record was unexpected because of the location of Buenos Aires. For a case where El Niño is known to cause readily observable changes of water level, consider the San Francisco sea level variations in Fig. 3.9. This figure, prepared in the same manner as Fig. 3.6, shows smoothed, detrended, and normalized values of monthly mean values of relative sea level (RSL) at San Francisco and the negative of the SOI as far back (to 1882) as available from the CPC web site. The correlation of San Francisco sea level anomalies with the SOI is significantly more striking than in the case of Buenos Aires. This is especially so after 1935. Trenberth (1997) has stated that SOI values prior to 1935 are reconstructions in many cases from barometric pressures other than from Tahiti and Darwin, and so may be a less accurate representation of ENSO. To the extent that U.S. West Coast sea level fluctuations reflect ENSO events, the high correlation ($r = 0.63$) of the SOI with San Francisco water level anomalies after 1935 bears out this contention. The correlation for the SOI and sea level anomalies prior to 1935 is $r = 0.43$.

The examples of San Francisco and Buenos Aires give clear warning that interannual and longer variations of sea level can have a very large effect on

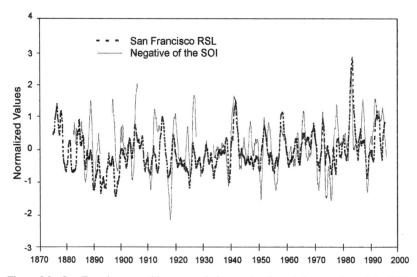

Figure 3.9 San Francisco monthly mean relative sea levels and the negative of the SOI.

trends computed even from long series. Figure 3.10 offers another view of the severity of this problem by displaying the trend of sea level at San Francisco for a 50-year sliding window. Each *point* in Fig. 3.10 is the trend of sea level in millimeters per year computed for the entire period of the preceding and following 25 years. Obviously an arbitrarily selected 50-year trend of sea level at San Francisco, which even changes *sign* at several points, does not reflect the underlying trend there! This example of dependence of sea level trend on epoch and record length, like the one at Buenos Aires, is larger than most, but not remarkably so.

What is revealed by these examples is that every sea level record must be examined, and a decision made as to whether the record is long enough to determine the underlying long-term trend of sea level. Some authors have used sea level records as short as 10 or 20 years in their analyses, which is in no case adequate. The most basic reason for the disagreements of authors concerning the rate of global sea level rise for the 20th century has come from a lack of a consensus on what constitutes an adequate length of sea level record. Increasingly, however, longer records are being used. Peltier and Tushingham (1989, 1991) in their analyses concluded that records of 50 years and longer gave an acceptable trade-off between adequate length and number of available records. Douglas (1991, 1992) argued that at least 60-year records were needed. Later Douglas (1997) found even better (i.e., more precise) results for global sea level rise by using records longer than 70 years. The case of Buenos Aires discussed here shows that even 80 years may not be long enough in some special cases unless identifiable extreme anomalies in

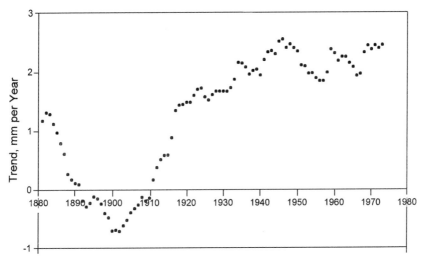

Figure 3.10 Fifty-year trends of sea level at San Francisco.

the sea level record can be identified and eliminated from the determination of the trend.

For a broader look at the role of record length in computed sea level trends, consider Fig. 3.11. It shows 20th-century trends of sea level obtained by linear regression for all tide gauge records from PSMSL RLR sites, plotted against record length. These trends have been corrected for GIA using the ICE-3G values of Peltier and Tushingham (1989). Vertical movement of the land due to GIA is a component of the sea level trend at any location. Unless a correction is applied, sea level trends determined from individual tide gauge records cannot be meaningfully compared, or aggregated to form an estimate of a global trend. Note that the longest records have GIA-corrected trends that converge to a positive value well in excess of 1 mm/yr with increasing record length. Figure 3.11 also shows clearly that estimates of sea level trends made from records as short as 10 or 20 years (e.g., Nakiboglu and Lambeck, 1991; Barnett, 1984; Emery and Aubrey, 1991) can be badly biased by the interannual-to-decadal variability water level.

There is some scatter visible in Fig. 3.11 even for trends determined from the longest record lengths. Most of this disparity arises from vertical crustal movements other than GIA and from data problems. The ubiquity of these issues concerning tide gauge records is illustrated by the redetermination of global sea level rise by Douglas (1997). There are about 175 RLR records >50 years in length but only 24 of these were used in this redetermination of global sea level rise; the others were rejected for geophysical and other reasons.

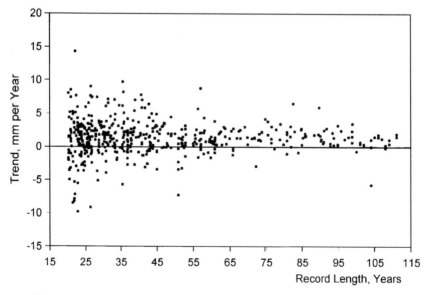

Figure 3.11 RLR-site sea level trends corrected for glacial isostatic adjustment.

3.4 TIDE GAUGE RECORDS SUITABLE FOR GLOBAL SEA LEVEL ANALYSES

The preceding section has shown how interannual and longer sea level variations create difficulties in determining underlying long-term trends of sea level. But determining an accurate trend is not enough, because even for very long records, the trend reflects processes other than a global increase of sea level. We turn now to the problem of identifying suitable long records from the very large data base of tide gauge data at the PSMSL. Figure 3.12 shows the most optimistic view of the situation. Plotted here are all of the PSMSL RLR tide gauge sites with records ≥20 years in length as of the 1997 data release. There are more than 500 sites on this map, but there is a very great location bias for the Northern Hemisphere and an absolute dominance of coastal positions rather than islands. (As an aside, this map illustrates why altimetric satellites such as TOPEX/Poseidon and JASON are required. There are simply not enough island sites to provide the necessary spatial coverage of the broad oceans for studies of basin-scale sea level variability and circulation.)

Figure 3.12 considerably overstates the number of potentially useful measuring sites (at least for the next 3 or 4 decades) because it includes tide gauge sites with records as short as 20 years. As the analysis of the previous section showed, much longer records are required to determine the underlying trend of sea level change at a site. Very long records go a long way toward relieving

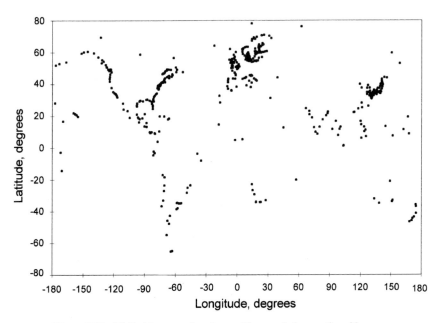

Figure 3.12 RLR tide gauge locations with records longer than 20 years.

the often-cited problem of the poor sampling of the oceans by tide gauges, at least at time scales approaching a century. We saw in Fig. 3.11 how long records gave GIA-corrected trends that converged toward a common value for the trend of sea level regardless of location. Groger and Plag (1993) and others have rightly pointed out the poor geographic distribution of tide gauge locations, but have failed to consider a critical factor. It is that the longer the period of a sea level variation, the greater the spatial extent of that signal. Thus very long records do not require the coverage needed by shorter ones to establish a value for global change. For the 20th century at least, Douglas (1991, 1992) has shown that widely separated tide gauge records show very consistent trends and (lack of) acceleration. Woodworth (1990) and Gornitz and Solow (1991) also observed the latter. Thus on a 100-year temporal scale, the lack of uniform global coverage appears to be much less important than usually supposed.

Sometimes tide gauge records may be unsuitable for mechanical or other causes affecting the sea level signal. This is where the use of multiple records from relatively nearby sites, as in the cases of San Francisco/San Diego and the U.S. middle Atlantic series previously discussed, can be decisive. The problem of the disagreement of sea level records at Sydney and Newcastle (Douglas, 1991) has already been mentioned; if this could be straightened out, a valuable long sea level record in the data-sparse Southern Hemisphere would be gained.

As an example of an obvious problem visible in a single record, consider Fig. 3.13 for the long series at Manila, the Philippines. The Manila record has

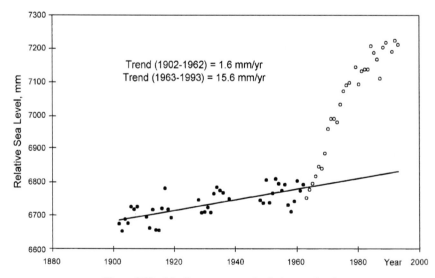

Figure 3.13 Manila mean annual relative sea level.

been used by many investigators of sea level rise, but obviously something happened in the early 1960s to cause an abrupt order-of-magnitude increase in the rate of rise there. The sudden increase is not plausibly oceanographic in origin. In the documentation accompanying this record, the PSMSL reports that harbor reclamation projects may be responsible. Whatever the reason, the Manila sea level record, although enticing for its length of nearly a century, is unsuitable for scientific use.

Sometimes problems with sea level time series have only a short duration, such as at Brest, France, during the World War II years. In this case, the anomalous values are recognizable by disagreement with the Newlyn, England, values, which otherwise are highly correlated (Douglas, 1992) over their common observation time. Subsequent to the problem with the Brest record, there is a gap of several years and then sea levels vary again in concert with those of Newlyn. In the documentation on their Web site, the PSMSL notes that gaps often are indicative of problems in series of sea level data and should be carefully investigated.

As important as data quality considerations may be, many high-quality and long tide gauge records for which an adequate GIA correction exists still show a large scatter in their trends. In most cases, plate tectonics is the reason. Douglas (1991) examined sea level trends from regional groupings of tide gauges that would be expected to show the same trend internal to each group based on oceanographic considerations and small differences of GIA. He found that Indian, Japanese, U.S. Pacific Northwest, and other locations associated with converging tectonic plates showed an inconsistency of sea level trends even over a small area. As an example, consider Fig. 3.14, which shows 5-year average relative sea levels for Seattle and Neah Bay, in Washington State in the United States. This is an area of converging tectonic plates, with associated seismicity and volcanism. The records are obviously morphologically similar at an interannual scale—note for example the usual U.S. west coast evidence of the 1982–1983 ENSO event. But the trends of sea level at these nearby sites are very different. Long-term relative sea level is increasing at Seattle and falling at Neah Bay. Geodesists have used the differing trends from these and other tide gauge records in the area to study vertical crustal movements, while at the same time the trends have been used by some investigators to study sea level rise. It should be clear by now that a tide gauge record cannot simultaneously serve geodesy and oceanography!

We turn now to the matter of providing a list of tide gauge records useful for determining a 20th-century rate of global sea level rise. From the preceding remarks and examples, it is clear that the tide gauge records must be long (>60–70 years at a minimum), be free of vertical crustal movements due to plate tectonics, be adequately correctable for GIA, and have trends insensitive to small changes in their record length or be capable of being edited in a justifiable manner based on oceanographic considerations. This is a long list

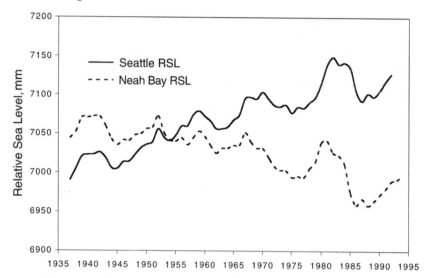

Figure 3.14   Relative annual sea level at Seattle and Neah Bay, Washington.

of restrictions, and only a small number of sites qualify. Table 3.2 gives the locations and trends of relative sea level for suitable 20th-century tide gauge records exceeding 70 years in length. Even though many of these records began before 1900, that date was selected as a starting point because of the evidence previously mentioned for a sharp increase in the rate of sea level in the mid-19th century, but without further significant acceleration over the entire 20th century (Douglas, 1992).

The groupings of the tide stations in Table 3.2 are presented according to oceanographic region. For example, U.S. east coast records are divided into sites north and south of Cape Hatteras, where the Gulf Stream leaves the coast and turns toward Europe. As remarked earlier, it is essential to group sea level trend measurements this way to prevent regions with a large number of gauges from having an excessive weight in the global solution.

The sea level trends in Table 3.2 have a significant scatter, but at least all are positive. In addition, they are not yet corrected for GIA. In Chapter 4 we shall see that applying a GIA correction, based on a new solution by W. R. Peltier, gives in general highly consistent results for sea level rise regardless of geographic location. However, in the Mediterranean, where the GIA correction is small (a few tenths of a millimeter per year), the sea level trends remain systematically small. Douglas (1997) noted that there has been no apparent increase of sea level at Marseille, Genova, and Trieste in the last 50 years, but could provide no explanation.

Table 3.2

20th-Century Relative Sea Level Trends from Records > 70 Years

Groups	Trend (mm/yr)	Start	End	Span (yr)
Aberdeen I+II	0.7	1900	1997	97
Newlyn	1.7	1915	1997	82
Brest	1.3	1900	1991	91
Cascais	1.6	1903	1991	88
Lagos	1.4	1909	1992	83
Marseille	1.2	1900	1996	96
Genova	1.2	1900	1992	92
Trieste	1.1	1905	1997	92
Auckland[1]	1.3	1904	1989	85
Dunedin[1]	1.4	1900	1989	89
Lyttelton[1]	2.3	1904	1989	85
Wellington[1]	1.7	1901	1988	87
Honolulu	1.5	1905	1997	92
San Francisco[2]	1.8	1900	1980	80
San Diego[2]	1.9	1906	1980	74
Balboa[2]	1.5	1908	1980	72
Buenos Aires[2]	1.1	1905	1980	75
Pensacola	2.1	1923	1996	73
Key West	2.2	1913	1997	84
Fernandina	2.0	1897	1993	96
Charleston	3.3	1922	1997	75
Baltimore	3.1	1903	1997	94
Atlantic City	4.0	1912	1997	85
New York	3.0	1900	1997	97
Boston	2.7	1921	1997	76
Portland	1.9	1912	1997	85
Halifax	3.4	1920	1996	76

[1] New Zealand trends from Hannah (1990).

[2] Data after 1980 eliminated to avoid effect of 1982/83 and 1997/98 ENSO events on trends.

3.5   ACCELERATION OF GLOBAL SEA LEVEL IN THE 20th CENTURY

It was mentioned in Chapter 2 that sea level began to rise at the modern rate approximately at the middle of the 19th century. In addition to the geological, geomorphic, and archeological evidence cited, there are supporting data from a few very long water level records. Woodworth (1999) has reconstructed some records reaching back to the early 18th century, which show a distinct increase in the rate of sea level rise in the 19th century. Figure 3.15 presents his data for Amsterdam sea levels. Similar results are seen at Brest, Liverpool, Sheerness, and Stockholm. Data taken prior to the late 19th century are from tide staff or other visual readings and have a significantly larger scatter of values than later data taken with a recording tide gauge. But the old data are adequate to show that there was a change in the rate of sea level rise near the middle of the 19th century. In Fig. 3.15 the data set was divided into two groups arbitrarily at 1850, but an alteration of a decade one way or the other would not change the results very much. The rate of relative sea level rise from 1850 onward is about 1.5 mm/ yr, about 5 times greater than prior to that time.

Has there been any acceleration of global sea level in the 20th century? Fortunately, global acceleration is much easier to address than global trend. This is so because GIA, and vertical crustal movements arising from plate tectonics at locations with long earthquake recurrence times, are both temporally linear over the tide gauge era. A polynomial regression of the 2nd degree in time can absorb the total linear component of the signal into the linear term whatever its source, and separately accommodate an acceleration in the quadratic term, if the record is long enough. Woodworth (1990), Gornitz and Solow (1991), and Douglas (1992) exploited this fact to examine the issue of an acceleration of sea level during the last 100–150 years. No author found conclusive evidence of a

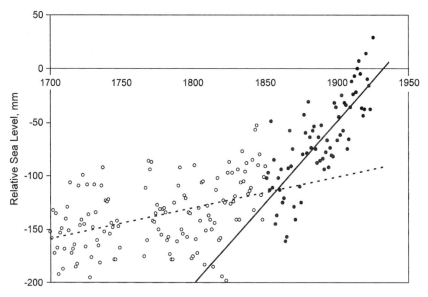

Figure 3.15   Amsterdam mean annual relative sea levels.

global acceleration of sea level, especially compared to what is predicted to accompany future global warming. Douglas (1992) used the largest number of tide gauge records for his estimate of global sea level acceleration and found for the period of the latter half of the 19th century to 1980 the global acceleration value -0.01 ± 0.01 mm/yr^2. He also concluded that 50 or more years would be needed to detect an increase in the rate of sea level rise (i.e., an acceleration) from ordinary tide gauge records. Figure 3.16 (after Douglas, 1992) illustrates the problem. It presents the coefficients in millimeters per year-squared of the acceleration terms derived by linear regression for PSMSL RLR sites with records >10 years, from 1880 to 1990. This figure is analogous to Fig. 3.11, which showed trends of sea level rise. However, in contrast to the trend case, it shows only a very small scatter of estimates for records longer than about 50–60 years. This demonstrates that vertical crustal movement rates are nearly constant at most sites; they are absorbed into the linear trends. The scatter of accelerations in Fig. 3.16 for the shorter records exists because interdecadal and longer variations of sea level, prevalent in sea level records, can be absorbed into the acceleration term for the shorter records. No reason is known for the precipitious drops in magnitude of the quadratic term at about 30 years and 60 years. In any case, Fig. 3.16 further underscores previous comments about the impact of interannual and longer variations of sea level, and the absolute necessity of using long records.

Concerning the issue of global warming and a possible increase in the rate of 21st-century sea level rise, Douglas (1992) has pointed out that predictions

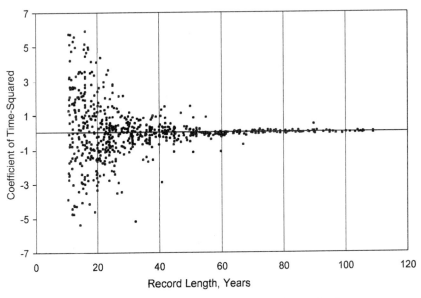

Figure 3.16   Acceleration component of relative sea levels.

(Warrick *et al.*, 1996) imply an acceleration of sea level corresponding to a coefficient of (time)^2 of order 0.1–0.2 mm/yr^2. This is small compared to the scatter of accelerations shown in Fig. 3.16 for sea level records less than about 60 years in length. However, Woodworth (1990) demonstrated by a numerical simulation that applying a meteorological correction for the long-period sea level variations at some sites may enable detection of a significant change in the rate of sea level rise in a few decades. Sturges and Hong in Chapter 7 of this book show explicitly how key U.S. east coast sea level records can be corrected to a very high degree for interdecadal variations of sea level. This gives some confidence that properly corrected tide gauge records do have the potential to detect a change of the underlying trend of regional sea level rise in a matter of decades.

3.6   CONCLUDING REMARKS

This chapter has shown that tide gauges record a (perhaps frustratingly) rich array of geophysical and oceanographic information. Vertical crustal movements and especially interannual and longer sea level variations make it difficult to obtain sea level trend estimates truly indicative of global sea level rise. The instrumental sea level record may not be stationary, nor does it consist of a simple trend plus measurement noise. These are problems that bedevil all climate records, as is well documented in National Research Council (1995). But even if these problems can be overcome to the satisfaction of researchers, major issues remain. What is needed for an understanding of global sea level and its relation to climate is an accurate budget of the contributors to sea level rise including thermal expansion and the various sources and sinks of water. As a practical matter for those hundreds of millions of persons living in threatened coastal regions and islands, the current rates of local (relative) sea level rise are already causing severe problems. Climate research and new observations should enable accurate predictions of future sea level rise to be made so that mitigation activities can be planned well in advance of disasters exacerbated by increased sea level, rather than in response to them.

REFERENCES

Barnett, T. P. (1984). The estimation of "global" sea level change: A problem of uniqueness. *J. Geophys. Res.* **89**, 7980–7988.

Barnett, T. P. (1990). Recent changes in sea level: A summary. In National Research Council *Sea-Level Change,* 37–51. National Academy Press, Washington, DC.

Chelton, D. B., and Davis, R. E. (1982). Monthly mean sea-level variability along the West Coast of North America. *J. Phys. Oceanog.* **12.**

Davis, J. L., and Mitrovica, J. X. (1996). Glacial isostatic adjustment and the anomalous tide gauge record of eastern North America. *Nature* **379.**

Davis, C. H., Kluever, C. A., and Haines, B. J. (1998). Elevation change of the southern Greenland ice sheet. *Science* **279**, 2086–2088.

Diaz, H. F., and Markgraf, V. (1992). *El Nino*. Cambridge University Press, Cambridge.

Douglas, B. C. (1991). Global sea level rise. *J. Geophys. Res.* **96**, no. c4, 6981–6992.

Douglas, B. C. (1992). Global sea level acceleration. *J. Geophys. Res.* **97**, no. c8, 12, 699–12, 706.

Douglas, B. C. (1995). Global sea level change: Determination and interpretation. *Revs. Geophys.* Suppl. 1425–1432.

Douglas, B. C. (1997). Global sea rise: A redetermination. *Surveys in Geophys.* **18**, 279–292.

Draper, N., and Smith, H. (1981). *Applied Regression Analysis*, 2nd ed. Wiley, New York.

Emery, K. O., and Aubrey, D. G. (1991). *Sea Levels, Land Levels, and Tide Gauges*. Springer-Verlag. New York.

Flemming, N. C. (1978). Holocene eustatic changes and coastal tectonics in the northeast Mediterranean: Implications for models of crustal consumption. *Phil. Trans. Royal Soc. London* **289**, 405–458.

Flemming, N. C. (1982). Multiple regression analysis of earth movements and eustatic sea-level change in the United Kingdom in the last 9000 years. *Proc. Geol. Assn.* **93**, 113–125.

Gornitz, V. (1995a). A comparison of differences between recent and late Holocene sea-level trends from eastern North America and other selected regions. *J. Coastal Res. Spec. Issue no.* **17**, 287–297.

Gornitz, V. (1995b). Monitoring sea level changes. *Climatic Change* **31**, 515–544.

Gornitz, V., and Solow, A. (1991). Observations of long-term tide-gauge records for indicators of accelerated sea level rise. In *Greenhouse Gas-Induced Climatic Change: A Critical Appraisal of Simulations and Observations,* Schlesinger, M. E. (ed.), Elsevier, Amsterdam, 347–367.

Groger, M., and Plag, H.-P. (1993). Estimations of a global sea level trend: Limitations from the structure of the PSMSL global sea level data set. *Global and Planetary Change* **8**, 161–179.

Hannah, J. (1990). Analysis of mean sea level data from New Zealand for the period 1899–1988. *J. Geophys. Res.* **95**, No. B8, 12, 399–12, 405.

Houghton, J. T., Meira Filho, L. G., Callander, B. A., Harris, N., Kattenberg, A., and Maskell, K. (eds.) (1996). *Climate Change 1995*, Intergovernmental Panel on Climate Change, Cambridge Univ. Press, Cambridge.

Maul, G. A., and Martin, D. M. (1993). Sea level rise at Key West, Florida, 1846–1992: America's longest instrument record? *Geophys. Res. Letters* **20**, 1955–1958.

Meier, M. F. (1984). Contribution of small glaciers to global sea level. *Science,* **226.**

Meier, M. F. (1990). Role of land ice in present and future sea level change. National Research Council, *Sea-Level Change,* National Academy Press, Washington, DC.

Mitrovica, J. X., and Davis, J. L. (1995). Present-day post-glacial sea level change far from the late Pleistocene ice sheets: Implications for recent analyses of tide gauge records. *Geophys. Res. Lett.* **22**, no. 18, 2529–2532.

Nakiboglu, S. M., and Lambeck, K. (1991). Secular sea level change. In *Glacial Isostasy, Sea Level, and Mantle Rheology,* Sabatini, R. Lambeck, K. and Boschi, E. (eds.), 237–258, Kluwer Academic, Dordrecht.

National Research Council (1995). *Natural Climate Variability on Decade-to-Century Time Scales*. National Academy Press, Washington, DC.

Peltier, W. R., and Tushingham, A. M. (1989). Global sea level rise and the greenhouse effect: Might they be connected? *Science* **244**, no. 4906, 806–810.

Peltier, W. R., and Tushingham, A. M. (1991). Influence of glacial isostatic adjustment on tide gauge measurements of secular sea level change. *J. Geophys. Res.* **96**, 6779–6796.

Peltier, W. R. (1996). Global sea level rise and glacial isostatic adjustment: An analysis of data from the East Coast of North America. *Geophys. Res. Lett.* **23**, no. 7, 717–720.

Peltier, W. R., and Jiang, Xianhua. (1997). Mantle viscosity, glacial isostatic adjustment and the eustatic level of the sea. *Surveys Geophys.* **18**, 239–277.

Pirazzoli, P. A. (1993). Global sea level changes and their measurement. *Global Planet. Change* **8**, 135–148.

Pugh, D. T. (1987). *Tides, Surges, and Mean Sea Level*. Wiley, New York.

Shennan, I., and Woodworth, P. L. (1992). A comparison of late Holocene and twentieth-century sea-level trends from the UK and North Sea Region. *Geophys. J. Int.* **109,** 96–105.

Trenberth, K. E. (1997). The definition of El Nino, *Bull. Am. Met. Society* **78,** no. 12, 2771–2777.

Trupin, A., and Wahr, J. (1990). Spectroscopic analysis of global tide gauge sea level data. *Geophys. J. Int.* **100,** 441–453.

Varekamp, J. C., Thomas, E., and van de Plassche, O. (1992). Relative sea level rise and climate change over the last 1500 years. *Terra Nova* **4,** 293–304.

Warrick, R. A., Le Provost, C., Meier, M. F., Oerlemans, J., and Woodworth, P. L. (1996). Changes in sea level. In *Climate Change 1995,* Intergovernmental Panel on Climate Change, Cambridge University Press, Cambridge.

Woodworth, P. L. (1990). A search for accelerations in records of European mean sea level. *Int. J. Climatology* **10,** 129–143.

Woodworth, P. L. (2000). High waters at liverpool since 1768: The UK's longest sea level record. *Geophysical Research Letters* **26,** No. 11, 1589–1592.

Wunsch, C. (1999). The interpretation of short climate records, with comments on the North Atlantic and southern oscillations. *Bull. Am. Met. Soc.* **80,** 245–255.

Zwally, H. J., Brenner, A. C., Major, A. J., Bindschadler, R. A., and Marsh, J. G. (1989). Growth of Greenland ice sheet: Measurement. *Science* **246,** 1587–1589.

Chapter 4 | Global Glacial Isostatic Adjustment and Modern Instrumental Records of Relative Sea Level History

W. R. Peltier

4.1 INTRODUCTION

In our effort to understand the implications of the secular variations of relative sea level so clearly recorded on tide gauges from which sufficiently long records are available (see Chapter 3), it is important that we recognize the extraordinary range of physical processes that contribute to such observations. Although it is now understood that the record length must be sufficiently long to allow us to average out the influence of the interannual variability associated with El Niño and other processes, it is not as widely understood that very much longer timescale geological processes may also significantly contaminate such observations. Here the word "contamination" refers to the contribution to the secular rate of change of sea level from any process other than those associated with modern climate drift, whether the latter be of "natural" or "anthropogenic" origin. This chapter focuses on the one specific source of geological timescale contamination that appears to dominate all others from the point of view of its global incidence. This concerns the phenomenon of glacial isostatic adjustment (hereafter GIA), a physical process caused by the intense cycle of glaciation and deglaciation to which the planet has been subjected for the past 900,000 years of the Pleistocene period of Earth history (e.g., see the discussion of the oxygen isotopic ice volume proxy from the ODP677 core in Shackleton *et al.,* 1990). Although other geological processes may also contribute to modern observations of secular rates of sea level change in specific regions, for example, near subduction zones where oceanic lithosphere descends into the mantle and where the descent may be accommodated seismogenically through the occurrence of intense dip–slip earthquakes such as are responsible for the continuing uplift of the coast of the Huon Peninsula of Papua New Guinea, these other processes seem not to be of global import and so will not be considered here.

To set the stage for the discussion to follow, consider the analysis of tide gauge secular sea level trends presented in Fig. 4.1, taken from Peltier and Tushingham (1989). It was first established in this paper that the glacial isostatic adjustment process was a significant global source of "contamination" of such

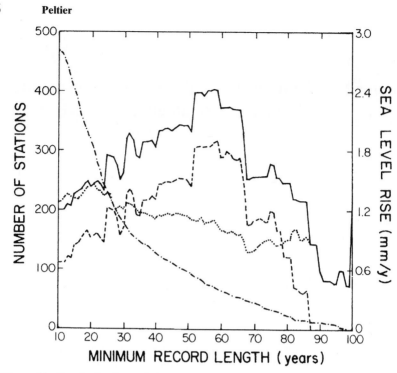

Figure 4.1 Results for the analysis of the global rate of RSL rise obtained from the linear regression analyses of Peltier and Tushingham (1989). The dash-dotted curve depicts the decrease in the number of tide gauge stations as the minimum record length employed in the analysis increases. The three individual estimates of the globally averaged rate of RSL rise versus minimum record length are based upon equal area averages of the raw (dashed curve) and GIA-corrected (solid curve) tide gauge data and a site-by-site average of the corrected data (dotted curve). The solid curve should provide the most accurate linear regression estimate of the present-day global rate of relative sea level rise.

observations. The figure shows (dash–dot line) the way in which the number of tide gauges decreases as a function of the length of the observational record that the individual instruments provide (as of the date of publication of the original paper). The same diagram also shows how the globally averaged rate of secular sea level change varies as a function of the lower timescale cutoff that is applied to determine whether the time series from a given gauge will be allowed to contribute to the determination of the globally averaged rate. The results obtained on the basis of such analyses are shown for two variants of the analysis procedure, one in which the "raw" rates of secular sea level rise are averaged by employing an areal weighting (dashed line) to account for the fact that the tide gauge installations do not sample the ocean surface uniformly, and one in which these areally weighted "raw" rates are "decontaminated" by subtraction of the influence on each gauge that is predicted by a theory of the GIA process (solid curve). In the third of the results (dotted curve) no attempt to weight the data for area coverage has been made; this is the least meaningful of the three estimates. Two points will be noted by inspection of these results. First, when the tide gauge (areally weighted) rates of secular sea level rise are corrected for the influence of glacial isostatic adjustment, the globally averaged rate of relative sea level (RSL) rise is somewhat increased. Second, and more important, however, is the fact that, as the minimum record length of the set of time series that is employed to determine the average is increased, the globally averaged rate of RSL rise also increases initially. This demonstrates clearly that, when short records are included in the set for which the global average is determined, the short timescale incoherence among them leads to an underestimate of the actual global rate of RSL rise that is ongoing in the Earth system. This point is further illustrated by Fig. 3.11 in the previous chapter of this book.

It is also notable, however, that in the analysis of Peltier and Tushingham (1989) the globally averaged rate determined on the basis of the longest records begins to fall rapidly once the minimum record length exceeds approximately 60 years. This fall is not compatible with the result shown in Fig. 3.11 in the preceding chapter. The reason for this has to do with the fact that in constructing Fig. 4.1 those records which were long but contaminated by local anthropogenic effects (ground water extraction, etc.) were not removed from the mix. To optimize our inference of the ongoing globally averaged rate of relative sea level rise, therefore, we will need to pay close attention to the application of three primary criteria. First, the records we employ must be sufficiently long to enable us to accurately average out the influence of the E1 Niño-related and other interannual variability. Second, we must eliminate all long records known to be strongly contaminated by well-recognized local tectonic or anthropogenic but not climate-related processes. Third, we need to filter these records to remove, as accurately as possible, the contribution to the local rate of secular sea level change at each tide gauge that is due to the influence of GIA.

In what is to follow, I will first review the global theoretical model of the GIA process that my students and I have developed over the past 20 years. Following a brief discussion of the global character of the solutions for the present day rate of RSL rise predicted by this theory, and of the quality of the fits to individual time series in the global data base of Holocene sea level histories that have been achieved with it, I will focus upon the results obtained by employing various versions of the global model to filter the same set of tide gauge data as were employed in Chapter 3 to infer the strength of the global signal that may be related to modern climate change. Finally, I will employ the theory in conjunction with specific observational constraints to show that the contribution to this inferred globally averaged rate of sea level rise that could be ascribed to persistent melting of the Antarctic and Greenland ice sheets since mid-Holocene time is extremely small. These results suffice to place the onus for the explanation of the observed modern rate of RSL rise squarely upon modern climate change.

4.2 THE MODERN GLOBAL THEORY OF THE GLACIAL ISOSTATIC ADJUSTMENT PROCESS

The theory of the GIA process is embodied in an integral "sea level equation," the solution of which describes the time-dependent separation between the surface of the solid earth and the gravitational equipotential surface that determines the equilibrium level of the sea. This equipotential surface is, of course, precisely the geoid of classical geodesy. The basic ingredients and initial form of this sea level equation (SLE) were developed in Peltier (1974, 1976), Peltier and Andrews (1976), and Farrell and Clark (1976). First solutions were published by Clark *et al.* (1978) and Peltier *et al.* (1978). This rather primitive initial theoretical structure has since been very significantly refined both in regard to the mathematical methods employed to solve the SLE and in regard to the range of physical effects included as influences upon the solution. The most general form of the SLE which governs the time dependence of relative sea level subsequent to the onset of any variation of land ice cover is as follows (see Peltier, 1998a for a most up-to-date review and detailed discussion), with $S(\theta, \lambda, t)$ the variation of relative sea level at latitude θ and longitude λ as a function of time t:

$$S(\theta, \lambda, t) = C(\theta, \lambda, t) \left[\int_{-\infty}^{t} dt' \iint_{\Omega} d\, \Omega' \, \{L(\theta', \lambda', t') \, G_{\phi}^{L} \, (\gamma, t - t') \right.$$
$$\left. + \, \Psi^{R} \, (\theta', \lambda', t') \, G_{\phi}^{T} \, (\gamma, t - t')\} + \frac{\Delta\Phi \, (t)}{g} \right]. \tag{1}$$

In Eq. (1) the function $C(\theta, \lambda, t)$ is the so-called "ocean function," which is, by definition, equal to unity where there is ocean and zero elsewhere. This equation is an integral equation because the surface mass load per unit area, L, is composed of both ocean–water and land–ice components. It takes the explicit form

$$L(\theta, \lambda, t) = \rho_{I} \, I(\theta, \lambda, t) + \rho_{w} \, S(\theta, \lambda, t) \tag{2}$$

in which $I(\theta, \lambda, t)$ is the space–time history of ice-thickness variations and $S(\theta, \lambda, t)$ is the space–time variation of ocean–water thickness variations. The latter is, of course, simply relative sea level itself. Due to the composite property of L expressed in Eq. (2), the unknown observable function S appears both on the left-hand side of (1) and under the triple convolution operator on the right-hand side of (1). The theory embodied in (1) is thus of integral equation form and may be employed to determine S given an assumed known history of glaciation and deglaciation $I(\theta, \lambda, t)$, together with suitable impulse response functions (Green functions) G_ϕ^L and G_ϕ^T. These Green functions embody all of the information concerning the radial viscoelastic structure of the earth that is required to determine the impact of this structure upon relative sea level history. In terms of the surface load and tidal "Love numbers," (h^L, k^L) and (h^T, k^T), respectively, these Green functions have the following mathematical forms, which consist of infinite series expansions in Legendre polynomials $P_\ell(\cos \theta)$ in which ℓ is spherical harmonic degree:

$$G_\phi^L (\theta, \lambda, t) = \frac{a}{m_e} \sum_{\ell=0}^{\infty} (1 + k_\ell^L (t) - h_\ell^L (t)) P_\ell (\cos \theta), \tag{3a}$$

$$G_\phi^T (\theta, \lambda, t) = \frac{1}{g} \sum_{\ell=0}^{\infty} (1 + k_\ell^T (t) - h_\ell^T (t)) P_\ell (\cos \theta). \tag{3b}$$

The Love numbers in turn have Dirichlet series expansions of the form (Peltier 1976)

$$h_\ell^L(t) = h_\ell^E \, \delta(t) + \sum_{j=1}^{J} r_j^\ell \, e^{-s_j^\ell t} \tag{4a}$$

$$k_\ell^L(t) = k_\ell^E \, \delta(t) + \sum_{j=1}^{J} q_j^\ell \, e^{-s_j^\ell t}. \tag{4b}$$

In these Dirichet series the $h_\ell^{H,E}$ and $k_\ell^{H,E}$ are precisely the elastic surface load Love numbers of Farrell (1972) which measure the magnitude of the response to application of a load that varies as a delta function in time. The quantity $\delta(t)$ is the Dirac delta function, the s_j^ℓ are a set of normal mode "poles" in the plane of the complex Laplace transform variable "s" representing the inverse relaxation times of modes of "viscous gravitational relaxation" (Peltier, 1976), and the r_j^ℓ and q_j^ℓ are the residues at these poles that may be determined in the usual way using Cauchy's residue theorem (Peltier, 1985) or the collocation technique described previously in Peltier (1974). The tidal Love number counterparts of (h_ℓ^L, k_ℓ^L), namely (h_ℓ^T, k_ℓ^T), have expansions of precisely the same form as (4a) and (4b) but they are determined subject to the boundary conditions that are appropriate when the responses arise through an imposed variation in the gravitational potential field rather than through an imposed surface mass load (see Farrell, 1972, for details).

Solutions to the sea level equation (1) are constructed using an iterative methodology that is based upon the recognition that the influence of time variations in the ocean function $C(\theta, \lambda, t)$ and the influence of the feedback onto sea level of the changing rotational state of the planet due to the variations of surface mass load described in (1) by the convolution of Ψ^R with G_ϕ^T are both second-order effects. The complete algorithm that I have developed to incorporate these effects, as discussed in Peltier (1994, 1998a, 1999), begins with the construction of a solution to (1) which neglects both of these effects and which is derived using the semispectral algorithm described in Mitrovica and Peltier (1991)—an algorithm that delivers a solution $S(\theta, \lambda, t)$ in terms of the time-dependent coefficients in an expansion of the solution on a basis of spherical harmonics. Solutions are now typically computed at very high spatial resolution by truncating the spherical harmonic expansions to degree and order 512. Given this first approximation to the full solution, we have complete knowledge of the variations of surface mass load associated with the glacial cycle, the component $I(\theta, \lambda, t)$ having been an assumed known input to (1) that may be varied to improve the fit to the observations and the component $S(\theta, \lambda, t)$ being the output from its solution. The surface load $L(\theta, \lambda, t)$ in (2) is thus entirely determined.

In the second step of this iterative process we next compute the rotational response to this history of surface loading by solving the Euler equation,

$$\frac{d}{dt}\left(J_{ij}\,\omega_j\right) + \varepsilon_{ijk}\,\omega_j\,J_{k\ell}\,\omega_k = 0, \tag{5}$$

in which J_{ij} is the moment of inertia tensor of the planet, ω_j are the components of its angular velocity vector, and ε_{ijk} is the Levi–Cevita alternating tensor. Assuming a biaxial form for the undeformed shaped of the planet, accurate solutions to (5) may be constructed by employing the standard perturbation expansion

$$\omega_i = \Omega\,(\delta_{ij} + m_i)$$

$$J_{ij} = I_{ij}, \qquad i \neq j$$

$$J_{11} = A + I_{11} \tag{6}$$

$$J_{22} = A + I_{22}$$

$$J_{33} = C + I_{33},$$

in which (A, A, C) are the principal moments of inertia, Ω is the basic state angular velocity of the earth, and I_{ij} and m_i are (assumed small) fluctuations away from the basic state values. To first order in these fluctuations (Munk and MacDonald, 1960), substitution of (6) into (5) leads to the following decoupled system for polar motion and earth rotation, respectively, as

$$\frac{i}{\rho_r}\,\dot{\mathbf{m}} + \mathbf{m} = \boldsymbol{\Psi} \tag{7a}$$

$$\dot{m}_3 = \dot{\Psi}_3, \tag{7b}$$

in which the overdot indicates a time derivative, the so-called excitation functions are $\boldsymbol{\Psi}$ and $\dot{\Psi}_3$, $\sigma_r = \Omega\,(C - A)/A$ is the Chandler wobble frequency of the rigid earth, $\mathbf{m} = m_1 + i\,m_2$, $\boldsymbol{\Psi} = \Psi_1 + i\,\Psi_2$, $i = \sqrt{-1}$, and the Ψ_i are

$$\Psi_1 = \frac{I_{13}}{(C - A)} + \frac{\dot{I}_{23}}{\Omega(C - A)}, \tag{8a}$$

$$\Psi_2 = \frac{I_{23}}{(C - A)} - \frac{\dot{I}_{13}}{\Omega(C - A)}, \tag{8b}$$

$$\Psi_3 = -\frac{I_{33}}{C}. \tag{8c}$$

Now the functions I_{ij} and thus the excitation functions Ψ_i are entirely determined by the surface loading history $L(\theta, \lambda, t)$ in Eq. (2), which is known from the first step in the iterative procedure [the details of these relationships are discussed in Peltier (1982) and Wu and Peltier (1984)]. Because the changes in the moments of inertia induced by surface loading also depend upon the redistribution of mass in the interior of the planet that is induced by the surface loading process, and because the redistribution of "internal mass" also depends upon the (assumed radial) viscoelastic structure, the excitation functions also depend upon this structure. Peltier (1982) and Wu and Peltier (1984) show in detail how to employ Laplace transform methods to construct complete solutions to Eqs. (7a) and (7b) for the time dependence of the perturbations to the angular velocity vector of the planet $\omega_i(t)$, $i = 1, 3$. No useful purpose would be served by repeating here the details of the algebraically complex analyses that are required to construct this solution for the rotational response to deglaciation.

In the third step of the iterative procedure we next construct the forcing function Ψ^R in (1) required to compute the impact upon sea level due to this changing rotational state. Because the solution of Peltier (1982) and Wu and Peltier (1984) used to determine the $\omega_i(t)$ is based upon the assumption that $\omega_i/\Omega \ll 1$, this same assumption must be made in computing Ψ^R. Dahlen (1976), in the context of his analysis of a different problem, has shown that the appropriate general solution for Ψ^R, correct to first order in perturbation theory, has the spherical harmonic expansion

$$\Psi^R = \Psi_{00} Y_{00}(\theta, \lambda) + \sum_{m=-1}^{+1} \Psi_{2m} Y_{2m}(\theta, \lambda), \tag{9}$$

in which the coefficients Ψ_{ij} are related to the $\omega_i(t)$ by

$$\Psi_{00} = \frac{2}{3}\omega_3(t)\,\Omega a^2 \tag{10a}$$

$$\Psi_{20} = -\frac{1}{3}\omega_3(t)\,\Omega a^2\,\sqrt{4/5} \tag{10b}$$

$$\Psi_{2-1} = (\omega_1(t) - i\omega_2(t))\,(\Omega a^2/2)\,\sqrt{2/15} \tag{10c}$$

$$\Psi_{2+1} = (\omega_1(t) + i\omega_2(t))\,(\Omega a^2/2)\,\sqrt{2/15}. \tag{10d}$$

Given Ψ^R we may now complete the third step in the iterative procedure by solving (1) once more for $S(\theta, \lambda, t)$, this time incorporating the influence of the changing rotational state of the planet, the first of the two second-order effects that the iterative procedure is intended to capture.

Now these first three steps in the iterative procedure have been based upon the assumption that the ocean function $C(\theta, \lambda, t)$ was constant and equal to its present-day form. The fourth step in the iterative procedure involves relaxing this assumption and this may be accomplished by following the procedure first described in Peltier (1994), which exploits the fact that the sea level equation (1), being itself a construct of first-order perturbation theory, delivers a solution for relative sea level $S(\theta, \lambda, t)$ that is "relative" to an unspecified and therefore arbitrary datum. This arbitrariness may be exploited to construct what I have previously referred to as a "topographically self-consistent" solution for RSL history. The procedure is simply as follows: First, we define a time-independent field $T'(\theta, \lambda)$ such that when we add it to the solution of the SLE computed for the present day (time $t = t_p$), we obtain the known present-day topography of the planet with respect to sea level, as

$$S(\theta, \lambda, t_p) + T'(\theta, \lambda) = T_p(\theta, \lambda), \tag{11}$$

in which T_p is the present-day topography wrt sea level, determined, say, by the ETOPO5 data set. Using (11) to define $T'(\theta, \lambda)$ we may then determine the topography at any time in the past by computing

$$T(\theta, \lambda, t) = S(\theta, \lambda, t) + [T_p(\theta, \lambda) - S(\theta, \lambda, t_p)]. \tag{12}$$

The true paleotopography, including the contribution from the ice sheets of thickness $I(\theta, \lambda, t)$, is then

$$PT(\theta, \lambda, t) = T(\theta, \lambda, t) + I(\theta, \lambda, t). \tag{13}$$

At any point in space and at any instant of time for which $PT < 0$ we therefore have ocean, whereas "land" is where $PT > 0$ (the only exception to this general rule concerns a number of low-lying inland seas (such as Caspian Sea) whose surfaces clearly do not lie on the same equipotential surface as does the global ocean). Subject to this caveat we may define a first approximation to the actual time-dependent ocean function as

$$C^1(\theta, \lambda, t) = 1 \quad \text{for} \quad PT < 0$$
$$C^1(\theta, \lambda, t) = 0 \quad \text{for} \quad PT > 0. \tag{14}$$

In reconstructing the topogarphy PT it is also important to recognize the "implicit" component of the ice load (Peltier 1998c) that is activated in the solution of (1) because of the perturbation theory–based nature of this equation. Since the fact that the redefinition of land to sea that occurs (for example) when the initially ice-covered Hudson Bay, Barents and Kara Seas, and the Gulf of Bothnia become connected to the ocean is a nonperturbative effect, the "explicit" component of the ice load that is varied in (1) to enable the model to fit the data must be considered to be an "effective" load. To compute the actual load and thus the true paleotopography, one must add to the effective load the additional "implicit" component. This completes step 4 of the iterative procedure. The next step in the process involves executing the first four steps in the process a second time, ending with a second guess for the time-dependent ocean function $C^2(\theta, \lambda, t)$ and, of course, with a new solution for relative sea level history $S(\theta, \lambda, t)$ that incorporates the full influence of rotational feedback.

Before discussing the impact of the global process of glacial isostatic adjustment on modern tide gauge–based estimates of secular sea level change, in the next section I will first discuss the global form of the solutions which the above-discussed theory delivers for the present-day rate of relative sea level rise and the extent to which it is able to provide satisfactory fits to a globally distributed set of relative sea level history observations. It will thereafter prove interesting not only to discuss the sea level signature that a specific tide gauge might be expected to measure but also to consider the signature of the time rate of change of absolute sea level (geoid height) that would be observed by an artificial Earth-orbiting satellite, either one equipped with an altimeter to measure distance from the satellite to the sea surface (e.g., TOPEX/POSEI-DON; see Chapter 6 for details of the way in which this system has already been employed to address the issue of global sea level change) or one that may be employed to infer this signal through measurement of the time dependence of the space-dependent strength of the gravitational field itself (expressed in terms of the time dependence of geoid height). The latter measurement is one that will be made at very high resolution in the near future using the GRACE satellite, and the CHAMP precursor, as discussed in greater detail in Peltier (1999).

4.3 GLOBAL PROPERTIES OF PRESENT-DAY SIGNATURES OF THE GIA PROCESS: TUNING THE MODEL

Since the solutions of the SLE that are of interest insofar as their predictions of present-day rates of sea level change are concerned are clearly those based upon models of the radial viscoelastic structure of the planet that best reconcile the available data base of geological timescale observations, it is clearly important to first establish that the theory embodied in the SLE does indeed satisfy these long timescale constraints when its physical properties are appropriately selected.

The primary data base used to tune the global model of the GIA process that is embodied in the SLE consists of radiocarbon-dated records of the variation of the level of the sea relative to the deforming surface of the solid earth. In tuning the properties of the global model to best fit these data, the elastic properties of the model of the planetary interior, namely, the density ρ and the two elastic Lamé parameters λ and μ, are assumed to be fixed, insofar as their variation with radius is concerned, by the results of free oscillation and body wave propagation seismology. I continue to base my analyses on an elastic structure defined by the PREM model of Dziewonski and Anderson (1981). Since the complete rheology of the interior is assumed to be adequately represented by a linear Maxwell solid, it is described by the Laplace transform domain stress-strain relation (see Peltier (1974) for discussion)

$$\tau_{ij}(s) = \lambda(s) \, e_{ij}(s) \, \delta_{ij} + 2\mu(s) \, e_{ij}(s), \tag{15}$$

in which s is the Laplace transform variable, τ_{ij} is the stress tensor, and e_{ij} is the strain tensor. The compliances $\lambda(s)$ and $\mu(s)$ of the linear Maxwell model are just

$$\lambda(s) = \frac{\lambda s + \mu \kappa / \nu}{s + \mu / \nu}, \tag{16a}$$

$$\mu(s) = \frac{\mu s}{s + \mu / \nu}. \tag{16b}$$

Since the bulk modulus $\kappa = \lambda + 2\mu/3 = \lambda(s) + 2\mu(s)/3$ is independent of s for this rheology, the model clearly has no bulk dissipation. The quantity $T_M = \nu/\mu$ is called the "Maxwell time" and this is the timescale that must be exceeded in order to effect the transition from Hookean elastic to Newtonian viscous behavior in response to an applied shear stress that is embodied in the Maxwell model. With (ρ, λ, μ) fixed by seismology, the only free parameter in the model is therefore the molecular viscosity ν. For the purpose of all of the analyses herein, the model of the interior of the earth is assumed to be spherically symmetric both in elasticity and in viscosity. It is this basic assumption that leads to the simple mathematical expressions (3a) and (3b) for the Green functions G_ϕ^L and G_ϕ^T. Since $\rho(r)$, $\mu(r)$, and $\lambda(r)$ of this spherically symmetric model are very well constrained seismologically (as defined by PREM), we are at liberty to vary only $v(r)$ in order to tune the model to fit the geological timescale observations. Given that the effective creep resistance of a solid (its viscosity) is strongly temperature dependent because the creep resistance of a solid is thermally activated, and because the mantle of the earth is assumed to be undergoing a process of thermally forced convection in which lateral variations of temperature are expected to be intense, it is clear that mantle viscosity must be a function of all three space coordinates. By employing a spherically symmetric model of the GIA process, we will therefore be testing the hypothesis that the aspherical viscosity structure that must characterize the actual mantle of the earth has no unambiguously observable effect. This is expected to be the case only if the actual lateral variations occur on sufficiently small horizontal scale, say over distances only on the order of the thickness of the lithosphere. As we will show immediately, spherically symmetric models of the internal structure can reconcile most of the available geological observations exceptionally well. There are, of course, exceptions to this general rule in the form of relative sea level observations from sites undergoing active tectonic uplift or subsidence, but these appear to constitute local exceptions to the general rule that the best fitting spherically symmetric model is able to reconcile the vast majority of the observations very well.

A sequence of three viscosity models for which I will explicitly discuss RSL predictions herein is illustrated in Fig. 4.2, where the models are labeled VM1, VM2, VM3 with the letters VM simply representing "viscosity model" and the number affixed following the letters being employed to distinguish the models from one another. Given a viscosity model and a model $I(\theta, \lambda, t)$ of the process of glaciation and deglaciation, we have all that is required to fix the inputs to Eq. (1) and thus to make predictions of postglacial relative sea level history. For all of the analyses discussed in this chapter, I will assume that the ICE-4G model described in Peltier (1994, 1996) adequately describes the last deglaciation even of the current ice age. Northern Hemisphere isopacks from this model for six different instants of time are shown in Fig. 4.3 (see color plate). It is important to understand that, although it is expected that this model is close in form to the actual deglaciation event that began approximately 21,000 calendar years ago, it cannot be exact and is still being refined in the course of work to develop an improved ICE-5G follow-on model.

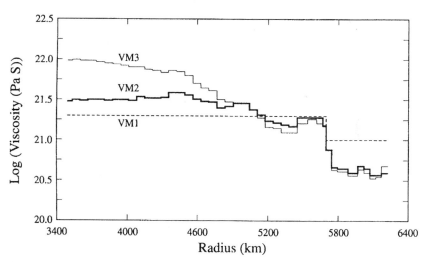

Figure 4.2   The VM1, VM2, and VM3 radial profiles of mantle viscosity discussed in the text. These represent the range of smooth viscosity profiles that have been employed for this physical property of the spherically symmetric viscoelastic models of the glacial isostatic adjustment process.

Figures 4.4 and 4.5 (see color plates) display comparisons of the present-day predicted rate of relative sea level rise as a function of the viscosity model (Fig. 4.4 shows the predictions of VM1 and VM2 together with their difference) and as a function of whether the influence of rotational feedback is included in the analysis (Fig. 4.5 presents results for viscosity model VM2 both including and excluding the rotational feedback effect as well as the difference between these two predictions). Inspection of these two figures will reveal all of the salient geographical characteristics of the global GIA process. Figure 4.4, for example, demonstrates that the regions that were ice-covered at LGM (Last Glacial Maximum) are regions in which relative sea level is currently falling as a consequence of ongoing postglacial rebound of the crust of the solid earth. Likewise, in the regions peripheral to those that were glaciated, relative sea level is predicted to be rising at present due to the collapse of the "glacial forebulge." The latter feature is a pronounced characteristic of the glaciated state and is a region in which the local radius of the planet is increased because of the mass that has been extruded from under the ice-loaded region as the earth isostatically adjusts to the weight of the surface load. When the ice load is removed and the rebound of the crust in the loaded region commences, the peripheral bulge of the solid earth begins to collapse. Beyond the region of forebulge collapse, in what I have previously referred to as the "far field" of the ice sheets, the rate of RSL change signal is characterized by a slow fall of RSL in ocean basin interiors and a region of similarly weak sea level rise confined to a "halo" around the coastline of each of the far field continents. The nature of the pattern of RSL rise/fall in the far field of the ice sheets may be understood as arising from the fact that, as the earth's shape relaxes following deglaciation, water is continuously "siphoned" from the central ocean basins in order to fill the depressions that are being created by proglacial forebulge collapse and by the "hydroisostatic tilting" of far field continental coastlines due to the weight applied to the surface by the offshore water load. Comparing the predictions of the VM1 and VM2 viscosity models demonstrates that the region in which the difference between them is largest is along the U.S. east coast. This region is therefore the one in which one would first look to discriminate between them.

Before discussing explicit analyses of this kind, it will be instructive to consider the similar set of predictions shown in Fig. 4.5, which has been designed to illustrate the impact of rotational feedback on the GIA-induced sea level signal. Results are shown in this figure for the VM2 model with and without this feedback included. The difference between the two predictions in the lower part of this figure shows that the influence of rotational feedback is very modest indeed, insofar as the amplitude of this signal is concerned. Furthermore, the pattern of the difference is almost entirely that of a degree 2 and order 1 Legendre polynomial, indicating on the basis of Eq. (10) that it is the polar wander component of the rotational response to GIA that feeds back most strongly onto sea level. It is important to note that the extremely weak influence of rotational feedback is at extreme odds with the contrary suggestion by Bills and James (1996), whose incorrect analysis of this problem led them to the conclusion that the strength of the rotational feedback was 1–2 orders of magnitude larger.

By way of demonstrating the way in which ^{14}C-dated RSL histories may be employed to discriminate between competing models of the radial viscosity structure, Fig. 4.6 shows an intercomparison of predictions of ICE-4G (VM1) and ICE-4G (VM2) for a sequence of 16 locations along the east coast of the continental United States at which ^{14}C-dated RSL histories are available. Also shown in this figure are RSL predictions for this set of locations for a model of Mitrovica and Forte (1997) which differs most extremely from both VM1 and VM2 in that it has a transition zone between 400 and 660 km depth in which the viscosity is extremely low (a feature that is balanced by the existence of a high-viscosity region in the middle of the lower mantle; see Peltier (1998a) for a figure explicitly comparing this model with VM2). Inspection of the results in Fig. 4.6 shows that the Mitrovica and Forte model is entirely excluded by the data because its "channel flow" characteristics predict the existence of raised beaches to the immediate south of the ice margin that are not observed. Channel flow models of this kind, in which the viscous flow of mantle material is confined to a low-viscosity near-surface region, have been known to be excluded by U.S. East coast RSL observations after the analyses presented by Cathles (1975) and Peltier (1974). Comparing the predictions of ICE-4G (VM1) and ICE-4G (VM2) demonstrates that at all sites along the northernmost part of the U.S. east coast the VM1-based model drastically overpredicts the present-day rate of RSL rise whereas the VM2 based model essentially eliminates these misfits.

As discussed most recently in Peltier (1998b), the viscosity model VM2 was in fact deduced on the basis of the application of the formal theory of Bayesian inference to a small subset of GIA data. In this procedure, model VM1 was employed as a starting model and the Fréchet kernels for the observed data were used to refine this model to optimally reduce the misfits of the starting model to the observations. None of the U.S. east coast data were employed in this process; rather they were withheld to provide a definitive check on the improvement to the model delivered by the automated inversion procedure. The data used for inversion actually consisted of the set of wavenumber-dependent relaxation times inferred by McConnell (1968) to characterize the postglacial recovery of Fennoscandia (the validity of which was recently reconfirmed by Weiczerkowski *et al.,* 1999), a set of approximately 23 relaxation times inferred on the basis of exponential fits to ^{14}C-dated RSL curves from both Canada and Fennoscandia, and the observed rate of nontidal acceleration of planetary rotation. VM2 serves as a good zeroth-order model of the radial variation of viscosity based upon numerous posteriori tests of its predictions against the very large set of observations that were not employed in its construction (see Peltier 1998a). Given this zeroth-order acceptable

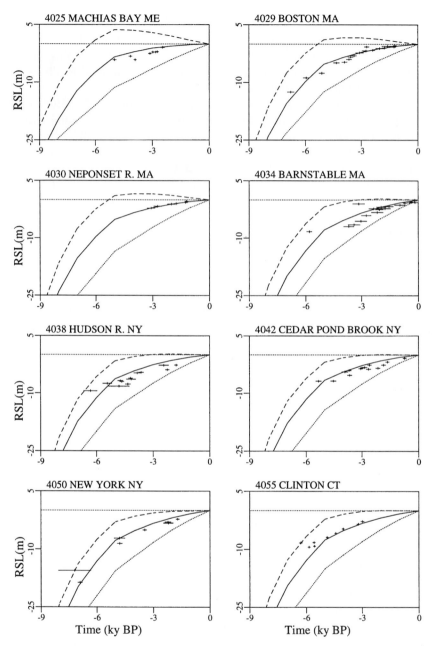

Figure 4.6 Predicted histories of relative sea level change at 16 sites along the eastern coast of the North American continent based upon use of the ICE-4G model of the deglaciation history. Results are shown for three different models of the radial variation of viscosity, respectively VM1 (dotted lines) and VM2 (solid lines) and a third model MF (dashed lines) recently obtained by Mitrovica and Forte (1997) by the simultaneous inversion of both glacial isostatic adjustment and aspherical geoid constraints. It will be clear that model MV2 entirely eliminates the misfits

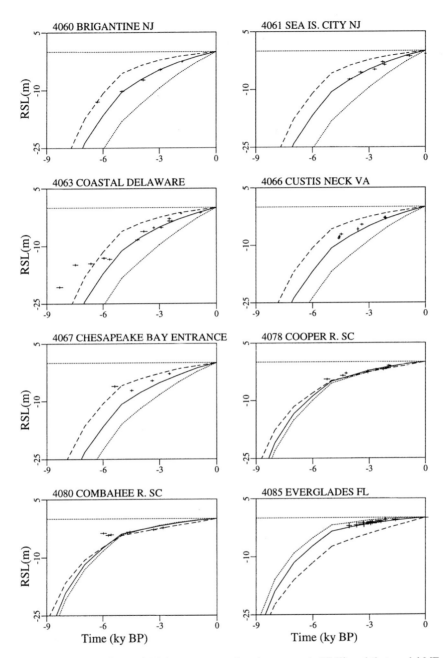

of model VM1 to these data (which were not employed to constrain VM2) and that model MF grossly misfits the observations at essentially all locations along this coast. As discussed in Peltier (1998a), model MF is a "channel flow" model in which the upper mantle and transition zone have such low viscosity that adjustment takes places through motions that are essentially confined to this region. Models of this kind have been known not to fit the GIA constraints since the earliest work of Cathles (1975) and Peltier (1974).

model for the global GIA process, we will proceed in the following section to examine the extent to which the GIA process is expected to contaminate tide gauge observations of secular sea level trends.

Before we pursue this line of argument, however, it will prove interesting to consider the nature of the global sea level signal that an artificial earth satellite designed to measure the time dependence of Earth's gravitational field might be expected to observe. Since this signal will be measured relative to the center of mass of the planet rather than relative to the surface of the solid earth, it will be clear that it need look nothing at all like the fields shown on Figs. 4.4 or 4.5. If we denote by $\dot{\mathrm{RSL}}$ (θ, λ, t) the rate of relative sea level change with respect to the surface of the solid earth, and by $\dot{\mathrm{RAD}}$ (θ, λ, t) the rate of change of the local radius of the planet measured with respect to the center of mass, then it will be clear that $\dot{\mathrm{ABS}}$ (θ, λ, t), the rate of change of absolute sea level measured with respect to the center of mass, is just

$$\dot{\mathrm{ABS}}\ (\theta,\ \lambda,\ t) = \dot{\mathrm{RSL}}\ (\theta,\ \lambda,\ t) + \dot{\mathrm{RAD}}\ (\theta,\ \lambda,\ t). \tag{17}$$

Figures 4.7 and 4.8 (see color plates) show both of the component parts of $\dot{\mathrm{ABS}}$ together with $\dot{\mathrm{ABS}}$ itself for models which respectively exclude (Fig. 4.7) and include (Fig. 4.8) the influence of rotational feedback, both of which employ the ICE-4G (VM2) model as input. Figure 4.9 (see color plate) shows the predictions with and without rotational feedback together with the difference between the predictions. The difference is clearly identical to the degree 2 and order 1 pattern shown previously in Fig. 4.5 as must, of course, be the case.

These results for absolute sea level, which constitute predictions of what the GRACE and CHAMP satellites should see if the GIA process were the only process currently influencing large-scale absolute sea level (geoid height) variability in the Earth System (see Peltier, 1999, for further discussion), demonstrate that the impact of rotational feedback on this observation will not be negligible even though it is negligible insofar as its impact upon sea level history relative to the surface of the solid earth is concerned. It will be extremely exciting over the next decade to keep careful watch on the outcomes of these ultra-high-resolution satellite experiments. They promise to provide a wealth of new knowledge concerning Earth system form and process. One useful prediction which we can make at present, however, by averaging the $\dot{\mathrm{ABS}}$ signals shown in Figs. 4.7 and 4.8 over the area of the oceans covered by TOPEX/POSEIDON, is that the global rate of RSL rise estimated on the basis of this measurement system must be corrected upward by approximately 0.30 mm/yr in order to correct for the ongoing influence of the glacial isostatic adjustment process. This is a significant upward revision of the number of 0.08 mm/yr previously cited in Peltier (1998) in reference to a first attempt by Rapp *et al.* (personal communication) to compute the magnitude of the GIA contamination based upon the ICE-4G (VM2) model predictions.

4.4   GIA-DECONTAMINATED TIDE GAUGE ESTIMATES OF THE RATE OF GLOBAL SEA LEVEL RISE

For the purpose of producing the best tide gauge–based estimate of the global rate of relative sea level rise which is not contaminated by the influence of the glacial isostatic adjustment process, I will focus first upon the results obtained by applying the highest resolution version of the ICE-4G (VM2) model for which results have been computed. These results are presented in Table 4.1 in which analyses are shown for 27 tide gauges, each of which is characterized by more than 70 years of observations. As discussed in detail by Douglas (1991), and as demonstrated by the earlier results of Peltier and Tushingham (1989) shown previously in Fig. 4.1, it is extremely important for the stability of the global estimate that one employ only very long records in such analyses since the interannual variability of relative sea level is intense and must be effectively averaged out if an accurate estimate is to be obtained. In Table 4.1 the individual columns give the latitude and longitude locations of each of the gauges, the number of years in the record and the estimate by Douglas ("Doug.") (Chapter 3) for the rate of relative sea level rise determined on the basis of a best fitting linear model to the raw annually averaged tide gauge data. Five columns of additional information labeled with the numbers −0.5, 0.5, 1.5, 2.5, 3.5 list the difference between the observed rates of RSL rise of Douglas and the GIA predictions for the ICE-4G (VM2) model of the rate of RSL rise that is predicted to occur at each of the tide gauge sites and at each of the times before present (measured in 1000 years) represented by the numbers −0.5 to +3.5, if the only process acting in the Earth System were the global process of glacial isostatic adjustment. The column labeled LSQ presents the difference between the Douglas calculation and the rate of relative sea level rise associated with the GIA process at each gauge determined by a linear least-squares fit to the theoretical predictions over the age range from 3.5 kyr before present to 0.5 kyr in the future. By including the LSQ result in our analysis, we will be able to estimate the magnitude of the error that would be committed by using the ^{14}C-dated geological data in the way employed by Gornitz (1995) in her analysis of U.S. east coast data. She obtained an estimate of the regional rate of RSL rise based upon GIA decontaminated tide gauge data from this coast of 1.5 mm/yr, a number that is significantly lower than that later obtained by Peltier (1996b), whose analysis of the data from the same region gave an average value near 1.9 mm/yr. At the bottom of the table is listed the average rate of RSL rise determined by simply averaging the observations of Douglas together with a standard deviation of the individual observations from this average value. For each of the

Table 4.1

Rates of Sea Level Rise, Adjusted from Tide Gauge Data

Station	LAT	LON	YRS	Doug. (mm/yr)	LSQ	Difts [Douglas–Peltier] (mm/yr)				
						−0.50	0.50	1.50	2.50	3.50
Lagos	37.10	−8.40	83	1.40	1.66	1.73	1.67	1.65	1.66	1.68
San Diego (Quara)	32.40	−117.00	74	1.90	1.78	1.98	1.85	1.76	1.71	1.68
Pensacola	30.20	−87.00	73	2.10	1.60	1.84	1.76	1.59	1.48	1.37
Fernandina	30.40	−81.30	96	2.00	1.38	1.60	1.59	1.38	1.25	1.07
Boston	42.20	−71.00	76	2.70	1.91	2.37	2.22	1.95	1.69	1.27
Halifax	44.40	−63.40	76	3.40	1.86	2.58	2.29	1.97	1.53	0.96
Aberdeen I & II	57.15	−2.08	97	0.70	1.20	1.18	1.12	1.19	1.24	1.33
Newlyn	50.10	−5.55	82	1.70	1.15	1.51	1.33	1.18	0.99	0.81
Brest	48.38	−4.50	91	1.30	0.73	1.10	0.92	0.76	0.57	0.37
Cascais	38.68	−9.42	88	1.60	1.84	1.91	1.84	1.82	1.84	1.86
Marseille	43.30	5.35	96	1.20	1.26	1.36	1.30	1.24	1.23	1.23
Genoa	44.40	8.90	92	1.20	1.30	1.38	1.32	1.29	1.28	1.30
Trieste	45.65	13.75	92	1.10	1.17	1.27	1.23	1.15	1.12	1.12
Auckland	−36.87	174.80	85	1.30	1.58	1.84	1.65	1.53	1.53	1.54
Dunedin	−45.88	170.50	89	1.40	1.79	1.93	2.00	1.68	1.67	1.68
Lyttelton	−43.60	172.72	85	2.30	2.68	2.88	2.70	2.65	2.65	2.68
Wellington	−41.28	174.78	87	1.70	2.10	2.33	2.16	2.06	2.04	2.06
Honolulu	21.32	−157.87	92	1.50	1.99	1.97	2.18	1.79	1.91	2.12
San Francisco	37.80	−122.47	80	1.80	1.52	1.50	1.63	1.52	1.44	1.35
Balboa	8.97	−79.57	72	1.50	1.74	1.80	1.74	1.68	1.73	1.83
Buenos Aires	−34.60	−58.37	75	1.10	1.59	2.17	1.98	1.35	1.36	1.41
Key West	24.55	−81.80	84	2.20	1.68	1.91	1.87	1.69	1.55	1.38
Charleston I	32.78	−79.93	75	3.30	2.47	2.86	2.74	2.51	2.33	1.85
Baltimore	39.27	−76.58	94	3.10	1.79	2.31	2.18	1.87	1.51	1.00
Atlantic City	39.35	−74.42	85	3.10	1.41	1.89	1.83	1.52	1.09	0.52
New York	40.70	−74.02	97	3.00	1.77	2.33	2.16	1.84	1.50	0.98
Portland	43.67	−70.25	85	1.90	1.83	2.07	2.02	1.85	1.70	1.46
Average				1.91	1.66	1.91	1.83	1.65	1.54	1.40
Standard deviation				0.75	0.40	0.47	0.45	0.41	0.41	0.50

five individual times for which the theoretical rates have been computed are listed the average differences between the Douglas observations and the GIA predictions together with a standard deviation. Also shown is the average difference between the observations and the LSQ estimate deduced by least-squares best fit of a straight line to the theoretical predictions over the age range from -3.5 kyr BP to $+0.5$ kyr in the future.

Careful inspection of the data presented in Table 4.1 indicates that there is no significant difference between the magnitude of the estimate of the globally averaged rate obtained by direct averaging of Douglas observations (this gives 1.91 mm/yr) and the result 1.87 mm/yr determined by averaging the GIA-corrected observations using the GIA prediction for the present day (by averaging the results for $+0.5$ and -0.5 kyr BP). There is, however, an extremely significant impact achieved by properly correcting the raw tide gauge data for the influence of GIA. Note that the standard deviation of the raw observations of Douglas from the average value is 0.75 mm/yr, whereas the standard deviation of the GIA-corrected rates from their average value is reduced to 0.45 mm/yr. It is therefore abundantly clear that when the tide gauge data are corrected for the GIA influence, the result is an estimate for the global rate of RSL rise that is much less variable as a function of position on Earth's surface. Also of considerable interest in this table is the difference between the globally averaged rates of relative sea level rise (a) determined by the GIA-corrected tide gauge data that are obtained when the GIA rates are accurate estimates for the present epoch and (b) those obtained when the GIA rate is taken to be represented by an average over the most recent 4 kyr of geological history. In the former case the globally averaged rate is 1.87 mm/yr, whereas in the latter case it is 1.66 mm/yr. The impact of employing the latter methodology of Gornitz (1995) is therefore severe, even when viewed from a global perspective. It inevitably leads to an underestimate of the average rate of RSL rise that could be associated with ongoing climate change in the Earth system. However, when viewed locally, as at the locations along the eastern seaboard of the continental United States, the impact of employing methodology (b) is further increased. On consideration of the results listed in Table 4.1 for the Key West, Charleston I, Baltimore, Atlantic City, and New York locations, application of method (b) for making the GIA correction, based upon the direct use of ^{14}C-controlled geological data averaged over the past 4 kyr to determine a rate, will reduce the strength of the inferred climate related contribution by amounts of 0.2, 0.32, 0.45, 0.44, 0.45 mm/yr for these five locations. The average value of the underestimate of the climate-related rate of RSL rise that is made by employing the Gornitz procedure at these five locations is therefore 0.37 mm/yr, which is essentially identical to the difference between the 1.5 mm/yr estimate of Gornitz (1995) and the 1.9 mm/yr estimate of Peltier (1996b) for the U.S. east coast region when the geological data are employed directly to decontaminate the observations. One must accurately estimate the GIA rate appropriate to the present day in order to accurately effect a decontamination of the tide gauge observations.

Several further consequential issues must be addressed before being satisfied that we have determined the best possible estimate of the global rate of RSL rise that could be associated with modern climate change. The first of these concerns the issue of the geographical representativeness of the raw average data presented in Table 4.1. There are clearly several regions of the Earth's surface that are oversampled by the analysis procedure employed in constructing Table 4.1. Specifically the east coast of North America is highly oversampled, there being six sites from the northern part of this coast (Baltimore, Atlantic City, New York, Boston, Portland, Halifax) and four from the southern part (Pensacola, Key West, Fernandina, Charleston I). Therefore 10 of the 27 gauges employed to produce a "global" estimate of the rate of RSL rise are actually located along one coast of the North Atlantic Ocean. To correct for this over sampling effect we choose to cluster the tide gauge sites as in Table 4.2 in order first to produce an average for each of the individual clusters and then to produce our estimate of the global rate of RSL rise by analyzing the average over these clusters. Inspection of the partition documented in Table 4.2 reveals the tide gauges have been amalgamated into 11 clusters, one of which consists of the data from Aberdeen I and II, both in Scotland and thus from a region that was heavily glaciated at Last Glacial Maximum. We will eliminate these data from further analysis for this reason. At the bottom of Table 4.2 we list the new globally averaged rate of RSL rise that is obtained by averaging over the individual cluster averages of the raw rates, which gives 1.71 mm/yr, and by averaging over the GIA-corrected cluster averages, which gives a value of 1.84 mm/yr. Of equal importance as the fact that the average rate is now somewhat increased when the GIA correction is made is the fact that the standard deviation of the individual cluster averages is reduced from 0.55 to 0.35 mm/yr. If we recall that the equivalent standard deviations for the unclustered data were 0.75 and 0.45 mm/yr, it will be clear that recognition of the oversampling problem has enabled us to further refine our estimate of the global rate of RSL rise that could be related to ongoing climate change in the Earth system.

We still need to understand the meaning of our best estimate, 1.84 mm/yr, of this average. What relative weight should be attached to the individual cluster averages in Table 4.2, considering the different fractional areas of Earth's surface that the individual clusters should represent? Because it would be arbitrary to assign a degree of importance to each of the clusters, I will leave this question as a caveat on the results that have been presented. It is worth noting, however, that the Northwestern European cluster represented by the tide gauges at Newlyn and Brest delivers results that are somewhat lower than those of most of the other clusters when the GIA correction is applied. This need not be taken to imply a problem with the analysis since there is no reason to believe that the influence of modern climate change

Table 4.2

Clustered Rates of Sea Level Rise, Adjusted from Tide Gauge Data

Station	LAT	LON	YRS	Doug. (mm/yr)	Diffs [Douglas–Peltier] (mm/yr) (Time = 0.00)
Aberdeen I & II	57.15	−2.08	97	0.70	1.15
1 location	Av rate of RSL rise, std dev = 0.70, 0.00; Av diff, std dev = 1.15, 0.00				
Newlyn	55.10	−5.55	82	1.70	1.42
Brest	48.38	−4.50	91	1.30	1.01
2 locations	Av rate of RSL rise, std dev = 1.50, 0.28; Av diff, std dev = 1.21, 0.29				
Cascais	38.68	−9.42	88	1.60	1.88
Lagos	37.10	−8.40	83	1.40	1.70
2 locations	Av rate of RSL rise, std dev = 1.50, 0.14; Av diff, std dev = 1.79, 0.12				
Marseille	43.30	5.35	96	1.20	1.33
Genova	44.40	8.90	92	1.20	1.35
Trieste	45.65	13.75	92	1.20	1.25
3 locations	Av rate of RSL rise, std dev = 1.17, 0.06; Av diff, std dev = 1.31, 0.05				
Auckland	−36.87	174.80	85	1.30	1.74
Dunedin	−45.88	170.50	89	1.40	1.97
Lyttelton	−43.60	172.72	85	2.30	2.79
Wellington	−41.28	174.78	87	1.70	2.24
4 locations	Av rate of RSL rise, std dev = 1.67, 0.45; Av diff, std dev = 2.19, 0.45				
Honolulu	21.32	−157.87	92	1.50	2.07
1 location	Av rate of RSL rise, std dev = 1.50, 0.00; Av diff, std dev = 2.07, 0.00				
San Francisco	37.80	−122.47	80	1.80	1.56
San Diego (Quara)	32.40	−117.10	74	1.90	1.91
2 locations	Av rate of RSL rise, std dev = 1.85, 0.07; Av diff, std dev = 1.74, 0.24				
Balboa	8.97	−79.57	72	1.50	1.77
1 location	Av rate of RSL rise, std dev = 1.50, 0.00; Av diff, std dev = 1.77, 0.00				
Buenos Aires	−34.60	−58.37	75	1.10	2.08
1 location	Av rate of RSL rise, std dev = 1.10, 0.00; Av diff, std dev = 2.08, 0.00				
Pensacola	30.20	−87.10	73	2.10	1.80
Key West	24.55	−81.80	84	2.20	1.89
Fernandina	30.40	−81.30	96	2.00	1.59
Charleston I	32.78	−79.93	75	3.30	2.80
4 locations	Av rate of RSL rise, std dev = 2.40, 0.61; Av diff, std dev = 2.02, 0.53				
Baltimore	39.27	−76.58	94	3.10	2.25
Atlantic City	39.35	−74.42	85	3.10	1.86
New York	40.70	−74.02	97	3.00	2.24
Boston	42.20	−71.00	76	2.70	2.29
Portland	43.67	−70.25	85	1.90	2.04
Halifax	44.40	−63.40	76	3.40	2.43
6 locations	Av rate of RSL rise, std dev = 2.87, 0.52; Av diff, std dev = 2.19, 0.20				

Average of group values
RSL rate from tide gauge records, std dev = 1.71, 0.55
RSL rate corrected for GIA, std dev = 1.84, 0.35

should cause sea level to rise at the same rate everywhere, especially if the global rise of level is caused to significant degree by the thermal expansion of the oceans as appears to be the case.

It will be noted on the basis of the data presented in both Tables 4.1 and 4.2 that the sign of the rate of RSL rise that is predicted to be occurring at all locations in the far field of the Pleistocene ice sheets is negative at all of the tide gauge locations from which long time series of data are available. Clearly when the individual tide gauge rates of RSL rise from this region are corrected by removing the GIA contamination, the result is an increase in the rate inferred to be due to ongoing climate change. As it happens the existence of this far field signature of the GIA process may be invoked to place strong constraints on the possible sources of the global rate of RSL rise that is presently occurring in the Earth system. The penultimate section of this chapter will address this interesting but insufficiently appreciated line of argument.

4.5 IMPLICATIONS OF THE MID-HOLOCENE HIGHSTAND OF SEA LEVEL AT SITES DISTANT FROM THE MAIN CONCENTRATIONS OF LGM LAND ICE

Figure 4.10 shows a location map for a large number of equatorial Pacific island locations from which relative sea level information is available for the Holocene epoch of Earth history. Figure 4.11 presents comparisons of the predictions of the same ICE-4G (VM2) model of the relative sea level history that should be observed at each of these island locations (shown as the solid line) together with the data available from each site as recently quality ensured in the paper of Grossman *et al.* (1998). From most of these oceanic islands there is usually only a single high-quality sea level datum available, often consisting of a ^{14}C-dated "notch" cut in coral that is currently observed to be located above local MSL. On each frame of this figure is also shown the predictions for two modifications of the Northern Hemisphere–constrained ICE-4G (VM2) model as well as for this base model itself. These modifications consist of adding to ICE-4G a continuous "melting tail" that continues to add meltwater to the oceans following the time at which the deglaciation event is assumed to have ended in ICE-4G, namely at 4 kyr BP. For the first of these modifications to the ICE-4G model, for which results are shown by the dotted lines on Fig. 4.11, the rate of eustatic sea level rise associated with late Holocene melting has been assumed to be equal to 0.25 mm/yr and all melting is assumed to derive from the Antarctic ice sheet. The second modification to ICE-4G, RSL results for which are shown as the dashed lines on Fig. 4.11, consists of a continuing eustatic sea level rise of strength 0.5 mm/yr, also assumed to be due to ongoing Antarctic melting.

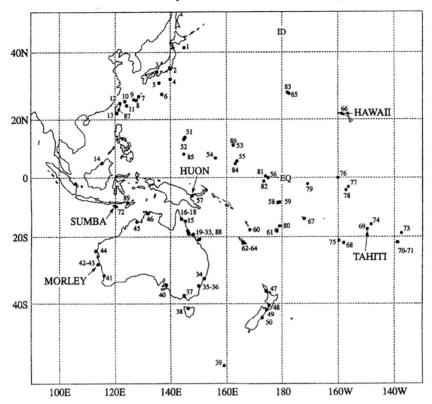

Figure 4.10   Location maps for equatorial Pacific Ocean island locations from which quality-controlled RSL constraints are available in the collection of Grossman *et al.* (1998) which overlaps the earlier compilation of Tushingham and Peltier (1992).

Inspection of the intercomparison of observations and theory shown in Fig. 4.11 demonstrates immediately that the strength of any late Holocene meltwater addition to the global oceans due to sustained melting of the Greenland ice sheet is strongly constrained by the elevation of the mid-Holocene highstand of sea level that is observed at essentially all islands in the equatorial Pacific Ocean from which actual shoreline elevation observations are available. This constraint is sufficiently tight that even continuous melting at the modest rate of 0.25 mm/yr leads to such a sharp reduction in the theoretically predicted amplitude of the highstand, to approximately 1 m rather than the observed 2 m, that it is ruled out by the observations. This same conclusion follows when the continuing melting of polar ice is assumed to derive from Greenland rather than from Antarctica (results not shown). With a rate of late Holocene melting of 0.5 mm/yr, the observed mid-Holocene highstand is entirely eliminated at all locations, as will be clear by inspection of Fig. 4.11. On this basis it would seem that the recent claim by Flemming *et al.* (1998) to the effect

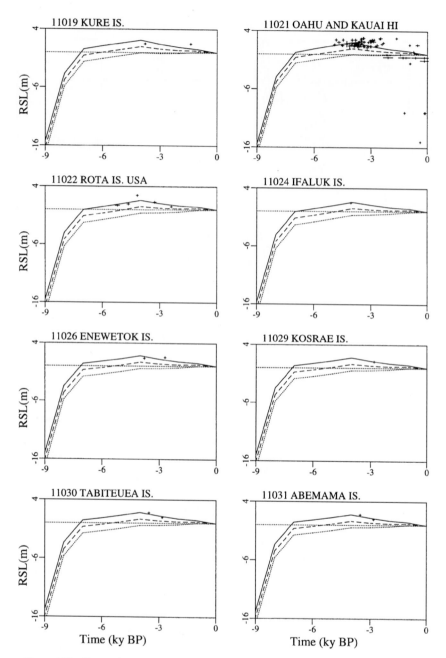

Figure 4.11 Comparisons of the predictions of the version of the gravitationally self-consistent theory of postglacial relative sea level history that neglects rotational feedback with the observations of many of the island locations shown in Fig. 4.10. All three of the time series shown at each site have been computed using the VM2 viscosity model but differ from one another in that they correspond to three variations on the ICE-4G deglaciation history. In particular the solid lines are the RSL predictions delivered by ICE-4G itself, the dotted lines are for a deglaciation

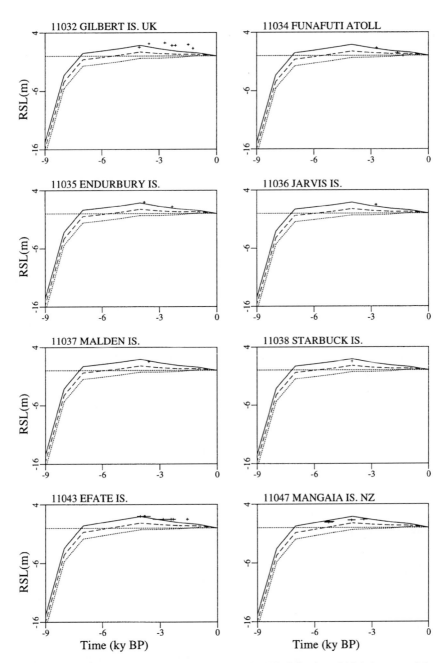

history that is identical to ICE-4G from LGM until 4 kyr BP, following which it is assumed that Antarctic melting continues to deliver water to the global ocean at an eustatic rate of 0.25 mm/ yr. The dashed RSL curves are for a model that is identical to that which produces the dotted RSL curves but for which the post–4-kyr BP "melting tail" is of strength 0.5 mm/yr. It will be noted that neither of the models that include late Holocene melting from Antarctica is able to fit any of the data.

that continuing late Holocene melting of the polar ice caps on Greenland and Antarctica might be contributing significantly to the tide gauge inferred modern rate of global sea level rise of 1.8 mm/yr, is untenable. They suggest that as much as 0.5 mm/yr of this signal could be due to such continuing polar ice sheet instability. Clearly this rate, or any significant fraction of it, would rule out the existence of the mid-Holocene highstand of sea level that is observed at both Pacific island locations and at virtually all coastal locations in the far field of the ice sheets (results not shown).

Since it is the existence of the GIA-related mid-Holocene highstand of sea level and the subsequent fall to modern levels that leads to the increase in the tide gauge rates of relative sea level rise at far field sites when the tide gauge rate is corrected for the influence of glacial isostatic adjustment (see the results for individual gauges shown in Table 4.1), we may also usefully express the impact of the late Holocene melting postulated by Flemming *et al.* (1998) in the form of global maps of the predicted present-day rate of RSL rise due to the GIA process. Figure 4.12 (see color plate) provides a useful demonstration of the utility of a presentation of this kind. The top of this figure shows the predicted present-day rate of RSL rise for the standard ICE-4G (VM2) model, whereas the middle and bottom respectively show equivalent predictions for the models that incorporate the influence of Antarctica derived "melting tails" of global strength 0.25 and 0.50 mm/yr. Clearly, when the strength of late Holocene melting is taken to be the larger value, the sign of the present-day rate of GIA-related RSL rise in the far field is no longer negative but is positive, as the highstand has been entirely eliminated.

On the basis of these analyses of the implications of the observed mid-Holocene highstand of sea level at sites distant from the polar ice sheets, it should be clear that the existence of this ubiquitous feature places strong constraints upon the extent to which these ice sheets may have continued to supply meltwater to the sea from mid-Holocene time onward. Based upon the analysis of these data discussed herein, I would therefore suggest that an upper bound on the rate of sustained mass loss from these systems should be taken to be no more than 0.1 mm/yr. This is clearly an insignificant fraction of the global rate of sea level rise of 1.8 mm/yr that we are required to explain. If the polar ice sheets are making a significant contribution to this global rate, then this contribution must have been activated only recently and could therefore plausibly be related to modern climate change. Even if we consider such contributions to be of relatively recent vintage, however, their magnitude is still rather strongly constrained by other geophysical observations. In particular, it has been demonstrated in Peltier (1998a, 1999) that Earth rotation observations of the nontidal acceleration of the rate of axial rotation and the speed and direction of true polar wander strongly constrain the present-day rate of mass loss from these polar locations. These analyses suggest that the maximum rate of collective mass loss from Greenland and Antarctica that could be accommodated by these observations is 0.5 mm/yr and this would be possible only under very special circumstances. In particular, Greenland would have to be the primary source of melting and the contributions from GIA and from the melting of small ice sheets and glaciers (e.g., Meier, 1984) would have to be most accommodating. If these constraints on the RSL contributions from the large polar ice sheets are correct, then we are clearly obliged to look elsewhere for explanation of the 1.8 mm/yr global sea level rise signal that the GIA-corrected tide gauge data reveal.

4.6   SUMMARY

A very detailed global viscoelastic theory of the process of glacial isostatic adjustment (GIA) has been developed. Application of this theory to the prediction of postglacial relative sea level histories has demonstrated that most ^{14}C-dated observations, from all sites in the global data base, are well explained by a spherically symmetric viscoelastic model whose elastic structure is fixed to that of PREM and whose radial viscosity profile is that of the VM2 model. Of course, there are exceptions to this general rule concerning the goodness of fit of the predictions of the spherically symmetric model to the observations. For example, at locations such as the Huon Peninsula of Papua New Guinea, where the entire coastline is being uplifted coseismically, the predictions of the GIA model fail to explain the observations (see Peltier, 1998a,d). It is expected that at other tectonically active locations similar misfits of the spherically symmetric theory to the observations should also be evident. Examples of such regions would certainly include the Mediterranean Sea region, Japan, and perhaps also the Pacific Northwest of North America where the Cordilleran ice sheet played a strong role in controlling the local history of relative sea level change but which is also influenced by active subduction.

These regions of misfit to the RSL predictions of the global viscoelastic theory of postglacial sea level change not withstanding, the extent to which this global spherically symmetric theory has been successful in reconciling the vast majority of the observations is satisfying, especially because only a very small subset of the observations has been employed to tune the model's radial profile of mantle viscosity. As discussed in greater detail in Peltier (1998b), these observations consisted of the set of wavenumber-dependent relaxation times determined by McConnell (1968) as characterizing the relaxation of Fennosandia following removal of its LGM ice load (the validity of which has recently be reconfirmed by Wieczerkowski et al., 1999, as previously mentioned), a set of 23 site-specific relaxation times from locations in both Canada and Fennoscandia, and the observed nontidal rate of the acceleration of axial rotation. The VM2 viscosity model that was determined on the basis solely of these data, using the formal procedure of Bayesian inference with the simple four-layer VM1 model as a starting model, was thereafter (Peltier

1996) shown to immediately reconcile the dramatic misfits of the starting model to the high-quality data set of ^{14}C-dated RSL histories that is available from the east coast of the continental United States (see also Peltier, 1998a). Because these data were not employed to constrain the radial viscosity structure, this is an extremely meaningful test of the validity of the model. That the new model also very well reconciles relative sea level data from far field sites throughout the equatorial Pacific Ocean has also been demonstrated explicitly in this chapter (see Figs. 4.9 and 4.10). The observations from the latter region offer a means by which we may strongly constrain the rate of mass loss from the great polar ice sheets on Antarctica and Greenland that may have been occurring continuously since mid-Holocene time. Our analysis demonstrates that the extent to which this influence could be contributing to the present-day observed rate of global sea level rise is negligibly small, a conclusion that is inconsistent with the claim to the contrary by Flemming *et al.* (1998).

Application of the global theory of the glacial isostatic adjustment process to filter this influence from the tide gauge data is clearly justified by the high-quality fits that the model delivers to the (widely distributed in space) observations of RSL variability on geological timescales over which ^{14}C dating may be employed to accurately determining sample age. As demonstrated through the analyses summarized in Tables 4.1 and 4.2, application of the GIA filter sharply reduces the standard deviation of the individual tide gauge measurements of the rate of RSL rise from their mean value, demonstrating the importance of this step in the analysis procedure. As demonstrated in Table 4.2, application of the filter to an aggregated set of tide gauge data, in which sites are lumped together if they are close in geographical location, also leads to an increase in the estimated global rate of RSL rise. In either case (Table 4.1 or Table 4.2) the best estimate we have been able to produce of the global rate of RSL rise that could be related to ongoing climate change in the Earth system is between 1.91 and 1.84 mm/yr.

An important additional result that follows from the results listed in Table 4.1 concerns the comparison between the GIA-corrected rates of RSL rise on tide gauges located along the east coast of the continental United States that would be obtained by least-squares fitting a straight line to the geological data over a period of 3–4 kyr and the result that is obtained by using the geological rate that obtains over the same time period over which RSL is sampled by the tide gauges. This has been investigated by using the GIA-predicted rates as proxy for the actual geological data and computing the GIA-corrected rates listed in the column labeled LSQ in Table 4.1. Comparing the results in this column with the average of those in the −0.5 and +0.5 kyr columns for all U.S. east coast sites will show that the procedure of least-squares fitting a straight line to the geological data over a period of 3–4 kyr will significantly overestimate the magnitude of the GIA-related signal and therefore its use will lead to a significant underestimate of the filtered tide gauge result. This fact very directly explains the reason for the approximately

0.4 mm/yr difference between the GIA-corrected rates for the U.S. east coast determined by Peltier (1996b) and those previously determined by Gornitz (1995), the former result being near 1.9 mm/yr and the latter near 1.5 mm/yr.

In concluding discussion of the analyses presented in this chapter, it is useful to reflect upon their implications concerning the relative importance of the various sources that might be contributing to the inferred global rate of relative sea level rise whose magnitude has been herein implied to be somewhat in excess of 1.8 mm/yr (between 1.91 and 1.84 mm/yr). The most recent estimates of the contribution from small ice sheets and glaciers (Meier and Bahr, 1996) are that this source has a strength of 0.3 ± 0.1 mm/yr. The influence of permafrost melting is expected to be even smaller with a strength of 0.1 ± 0.1 mm/yr. I have argued herein that the contribution due to continuing late Holocene melting of polar ice from either Antarctica or Greenland is bounded above by 0.1 mm/yr. Since the most recent estimate of the terrestrial storage term (Chapter 5) suggests this to be -0.9 ± 0.5 mm/yr (note that this is revised from the previous estimate of -0.3 ± 0.15 mm/yr obtained by Gornitz *et al.* 1997) there is clearly a residual that requires explanation in terms of significant contributions from either Greenland and/or Antarctica and/or from the thermal expansion of the oceans. Since the geophysical constraint through Earth rotation observations (Peltier 1998a, 1999) appears to require the former to be less than 0.5 mm/yr, the implication of these arguments would appear to be that the current rate of global sea level rise due to thermal expansion of the oceans might be significantly larger than the rate usually assumed to best represent this contribution (0.6 ± 0.2 mm/yr). In connection with the latter contribution, however, it is not at all clear that the current generation of coupled atmosphere–ocean models, the results from which provide a primary basis for this estimate, are capable of accurately gauging the significance of this steric effect. Clearly much further effort, especially in strengthening the observational constraint on the steric signal and in more precisely estimating the contribution due to terrestrial storage will be required before we shall be in any position to be confident as to which of these conventionally considered influences is more important. If terrestrial storage were entirely unimportant, then the observed present day rate of rsl rise would be within the upper bound defined by the net influence of the other contributions. However, if the (negative) influence of terrestrial storage is as large as the most recent estimate (see Chapter 5), then the influence of thermal expansion (or one of the other contributions) would have to be considerably larger than the above stated estimates in order that the inferred global rate of rsl rise be successfully explained.

REFERENCES

Bills, G. B. and James, T. S. (1996). Late Quaternary variations in relative sea level due to glacial cycle polar wander. *Geophys. Res. Lett.* **23**, 3023–3026.

Cathles, L. M. (1975). *The Viscosity of the Earth's Mantle.* Princeton Univ. Press, Princeton, NJ.

Clark, J. A., Farrell, W. E., and Peltier, W. R. (1978). Global changes in postglacial sea level: A numerical calculation. *Quat. Res.* **9,** 265–287.

Dahlen, F. A. (1976). The passive influence of the oceans upon the rotation of the Earth. *Geophys. J. R. Astron. Soc.* **46,** 363–406.

Douglas, B. D. (1991). Global sea level rise. *J. Geophys. Res.* **96,** 6981–6992.

Dziewonski, A. M. and Anderson, D. L. (1981). Preliminary reference Earth model. *Phys. Earth Planet. Inter.* **25,** 297–356.

Farrell, W. E. (1972). Deformation of the Earth by surface loads. *Rev. Geophys.* **10,** 761–797.

Farrell, W. E. and Clarke, J. A. (1976). On postglacial sea level. *Geophys. J. R. Astron. Soc.* **46,** 647–667.

Fleming, K., Johnston, P., Zwartz, D., Yokoyama, Y., Lambeck, K., and Chappell, J. (1998). Refining the eustatic sea level curve since the Last Glacial Maximum using far-intermediate-field sites. *Earth Planet. Sci. Lett.* **163,** 327–342.

Gornitz, V. (1995). A comparison of differences between recent and late Holocene sea-level trends from eastern North America and other selected regions. *J. Coastal Res.* **17,** 287–297.

Gornitz, V., Rosenzweig, C., and Hillel, D. (1997). Effects of anthropogenic intervention in the land hydrological cycle on global sea level rise. *Global and Planetary Change* **14,** 147–161.

Grossman, E. E., Fletcher, C. H., III, and Richmond, B. M. (1998). The Holocene sea-level high stand in the equatorial Pacific: Analysis of the insular paleosea-level data base. *Coral Reefs* **17,** 309–327.

McConnell, R. K. (1968). Viscosity of the mantle from relaxation time spectra of isostatic adjustment. *J. Geophys. Res.* **73,** 7089–7105.

Meier, M. (1984). Contribution of small glaciers to global sea level. *Science* **226,** 1418–1421.

Meier, M. F. and Bahr, D. B. Counting glaciers: Use of scaling methods to estimate the number and size distribution of the glaciers of the world. In Colbeck, S. C. (ed.), *Glaciers, Ice Sheets and Volcanoes: A Tribute to Mark F. Meier.* CRREL Special Report 96-27, 89-94. U.S. Army, Hanover, NH.

Mitrovica, J. X., and Peltier, W. R. (1991). On postglacial geoid subsidence over the equatorial oceans. *J. Geophys. Res.* **96,** 20053–20071.

Mitrovica, J. X. and Forte, A. M. (1997). Radial profile of mantle viscosity: Results from the joint inversion of convection and postglacial rebound observables. *Geophys. J. Int.* **102,** 2751–2769. 1997.

Munk, W. H. and MacDonald, G. F. (1960). *The Rotation of the Earth.* Cambridge Univ. Press, New York, 1960.

Peltier, W. R. (1974). The impulse response of a Maxwell Earth. *Rev. Geophys.* **12,** 649–669.

Peltier, W. R. (1976). Glacial isostatic adjustment II: The inverse problem. *Geophys. J. R. Astron. Soc.* **46,** 669–706.

Peltier, W. R. (1982). Dynamics of the ice-age Earth. *Adv. Geophys.* **24,** 1–146.

Peltier, W. R. (1985). The LAGEOS constraint on deep mantle viscosity: Results from a new normal mode method for the inversion of viscoelastic relaxation spectra. *J. Geophys. Res.* **90,** 9411–9421.

Peltier, W. R. (1994). Ice-age paleotopography. *Science* **265,** 195–201.

Peltier, W. R. (1996a). Mantle viscosity and ice-age ice-sheet topography. *Science* **273,** 1359–1364.

Peltier, W. R. (1996b). Global sea level rise and glacial isostatic adjustment: An analysis of data from the east coast of North America. *Geophys. Res. Lett.* **23,** 717–720.

Peltier, W. R. (1998a). Postglacial variations in the level of the sea: Implications for climate dynamics and solid earth geophysics. *Rev. Geophys.* **36,** 603–689.

Peltier, W. R. (1998b). The inverse problem for mantle viscosity. *Inverse Problems* **14,** 441–478.

Peltier, W. R. (1998c). "Implicit Ice" in the global theory of glacial isostatic adjustment. *Geophys. Res. Lett.* **25,** 3957–3960.

Peltier, W. R. (1998d). Glacial isostatic adjustment and coastal tectonics. In *Coastal Tectonics, Special Publication No. 146*, Geological Society of London, 1–30.

Peltier, W. R. (1999). Global sea level rise and glacial isostatic adjustment. *Global and Planetary Change* **20**, 93–123.

Peltier, W. R., and Andrews, J. T. (1976). Glacial isostatic adjustment. I. The forward problem. *Geophys. J. R. Astron. Soc.* **46**, 605–646.

Peltier, W. R., Farrell, W. E., and Clark, J. A. (1978). Glacial isostasy and relative sea level: A global finite element model. *Tectonophysics* **50**, 81–110.

Peltier, W. R., and Tushingham, A. M. (1989). Global sea level rise and the greenhouse effect: Might they be connected? *Science* **244**, 806–810.

Shackleton, N. J., Berger, A., and Peltier, W. R. (1990). An alternative astronomical calibration of the lower Pleistocene timescale based upon ODP site 677. *Trans. R. Soc. Edinburgh Earth Sci.* **81**, 251–261.

Wieczerkowski, Karin, Mitrovica, Jerry X., and Wolf, Detlef. (1999). A revised relaxation-time spectrum for Fennoscandia. *Geophys. J. Int.* **139**, 69–86, 1999.

Wu, P. and Peltier, W. R. (1984). Pleistocene deglaciation and the Earth's rotation: A new analysis. *Geophys. J. R. Astron. Soc.* **76**, 202–242.

| Chapter 5 | Impoundment, Groundwater Mining, and Other Hydrologic Transformations: Impacts on Global Sea Level Rise |

Vivien Gornitz

5.1 INTRODUCTION

The 20th century has seen a rise in the world's mean sea level of about 10–25 cm, with a most likely value of 18 cm (Gornitz, 1995; Warrick et al., 1996; Chapter 4, this book). Rates of sea level rise (SLR) may increase by 2–5 times over present rates within the next 100 years due to projected global warming, posing a threat to low-lying coastal regions (Chapter 8). Nearly half of the observed SLR of 18 cm during the 20th century can be attributed to thermal expansion of the upper ocean and melting of mountain glaciers associated with rising global mean surface air temperature. Greater uncertainty surrounds the contribution to SLR from melting of the Greenland and Antarctic ice sheets, which may be close to zero at present (Table 5.1). Thus, the balance of 20th century SLR beyond that attributable to ocean thermal expansion and melting of small glaciers needs to be accounted for by other processes, including human-induced transformations of the terrestrial hydrologic cycle.

It has long been recognized that anthropogenic activities[1] could influence stream runoff, and hence ultimately affect global sea level (Newman and Fairbridge, 1986; Gornitz and Lebedeff, 1987; Chao, 1991; 1994; 1995; Sahagian et al., 1994; Rodenburg, 1994; Gornitz et al., 1997). However, quantitative assessments have been lacking. The Intergovernmental Panel on Climate Change, in reviewing the literature, estimated that these actions contribute between −0.4 to +0.75 mm/yr to sea level (Table 5.1; Warrick et al., 1996). This considerable spread points to the large uncertainty that exists over the magnitude and even the sign of such anthropogenic effects on global sea level. In this chapter, we review the important hydrologic processes and present historical data pertaining to human modification of the land's hydrologic system since the beginning of this century. We also briefly examine improved hydrologic models that consider land cover transformations, and new, promising remote sensing techniques to monitor hydrological impacts.

[1] These include groundwater mining, impoundment in reservoirs, vegetation clearance, infiltration beneath reservoirs and irrigated fields, and evaporation from reservoirs and irrigated fields.

Table 5.1

Estimated Contributions to Global Sea Level Rise during the Past 100 Years (cm)
(after Warrick *et al.*, 1996)

Component	Low	Medium	High
Thermal expansion	2	4	7
Mountain glaciers	2	3.5	5
Greenland ice sheet	− 4	0	4
Antarctic ice	−14	0	14
Impoundment and groundwater mining	−5	0.5	7
Total	−19	8	37
Observed	10	18	25

Source: Intergovernmental Panel on Climate Change.

5.1.1   An Overview of the Land Hydrologic Cycle

The earth's water resides in four major reservoirs: the ocean, glaciers and ice sheets, terrestrial water, and the atmosphere (Fig. 5.1). The oceans contain 96.5% of the total water reserves, whereas freshwater reserves constitute only 2.5% of the total (Fig. 5.1). Around 69% of freshwater is locked up in glaciers, permafrost, and ice, while another 30% lies in groundwater and soil, leaving less than 1% in lakes and rivers (Jones, 1997; Shiklomanov, 1997). Water is exchanged between these reservoirs through the processes of evaporation, precipitation, and runoff. The net balance between the components of the hydrologic system is expressed as

$$P - E \pm S = R, \tag{1}$$

where P is precipitation, E is evaporation/evapotranspiration, S is change in storage (soils, rocks, lakes), and R is stream runoff. This equation can be

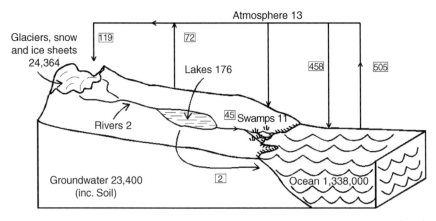

Figure 5.1   The hydrologic cycle. Figures (after Shiklomanov, 1997) are in units of 10^3 km^3.

solved for any time period, but in practice, an annual or flood hydrograph period is selected, such that initial and final conditions are reasonably equivalent.

Around 500,000 km^3 of water evaporates annually from the world's oceans, nearly 90% of which precipitates over the ocean and 10% over land (Shiklomanov, 1993). Ocean-derived moisture together with local land sources accounts for terrestrial precipitation of around 119,000 km^3/yr. Of this amount, between 40,700 and 47,000 km^3/yr returns to the sea as total runoff (Fig. 5.1; Shiklomanov, 1993; L'vovich and White, 1990). Over half of the runoff comes from the major southeast Asian rivers draining the Himalayas and the Amazon River in South America. Analysis of historical records of 142 rivers throughout the world with over 50 years of historical data, occupying drainage areas greater than 1000 km^2, revealed no consistent trends in streamflow over time (Houghton et al., 1996). On the other hand, large-scale interdecadal variations correlate well with precipitation in certain regions (e.g., Marengo et al., 1994).

A fraction of the water falling over dry soil infiltrates into the ground, at rates depending on the physical properties of the soil and its water content. Water on the surface may evaporate, or, if the soil is saturated, flow downslope as surface runoff into streams and lakes, or slowly percolate toward the water table[2] and recharge groundwater. The exact volume of water held in groundwater is uncertain, but exchanges between relatively shallow groundwater stores may occur over tens to hundreds of years, whereas exchanges at deeper levels may take thousands of years or more (Jones, 1997). For example, water in some Saharan aquifers may date back to the wetter, "pluvial" climate following the end of the last glacial period.

Vegetation participates in the hydrological cycle by modifying evaporation and runoff, compared with bare soil. Over vegetated terrain, the plant canopy intercepts a fraction of the falling water, the throughfall reaches the soil and infiltrates into the ground, where plant roots draw water from the soil. Leaves evapotranspire water mainly during the day, driven by solar heating and photosynthesis. Land-use changes, such as clearing of the natural vegetation for agriculture, grazing, or urbanization, in particular impact hydrological interactions among vegetation, soil and atmosphere, such as interception, infiltration, percolation, and evapotranspiration (Table 5.2). Some of these impacts will be discussed in greater detail below.

Within the last 50–100 years, human appropriations of water, although still a relatively small fraction of the total volume of water exchanged, have nonetheless become significant. For example, Postel et al. (1996) estimate that people already utilize 54% of readily accessible river runoff. River runoff is increasingly being managed, by construction of dams for hydroelectric power, flood control, and irrigation, as well as by huge intrabasin transfers of water,

---

[2] The surface separating the saturated zone (i.e., the zone where all pore spaces in the soil, sediment, or rock are filled with water) from the unsaturated zone where pore spaces are occupied by air and some water.

Table 5.2

Hydrological Impacts of Land-Use Change

Land-use change	Hydrological effects
Deforestation	Decreases in interception, evapotranspiration, infiltration capacity, soil porosity; increases in surface runoff, annual flows, flooding; snowmelt hastened; changes in seasonal distribution of runoff.
Urbanization	Decreases in infiltration and percolation, increases in surface runoff, changes in groundwater levels.
Agriculture	Decreases in infiltration, increases in surface runoff, changes in evapotranspiration, changes in seasonal distribution of runoff.
Wetlands drainage	Decreases in water retention capacity, evapotranspiration; increases in floods, annual flows.

and channelization of rivers. Other changes in runoff accompany alteration of land use or land cover patterns, such as replacement of forests by grassland or cropland, urbanization, or draining and clearance of wetlands (Table 5.2). Good summaries and case studies of these processes and their consequences can be found in Walling (1987), Newson (1992), Jones (1997), L'vovich and White (1990), Gleick (1993), and Shiklomanov (1993, 1997). The major ways in which human transformation of the hydrologic cycle could influence sea level are illustrated schematically in Fig. 5.2 (see color plate).

5.1.2   Anthropogenic Changes in Runoff and Sea Level Rise

The contribution to sea level rise (SLR_a) of anthropogenic interventions in the land hydrologic cycle, over a specified time period, can be expressed as

$$SLR_a = (G + U + C + D + W) - (R + I), \qquad (2)$$

where G = SLR due to groundwater mining
  U = SLR due to increased runoff from urbanization
  C = SLR due to water release from combustion of fossil fuels and decomposition of biomass and soil organic matter
  D = SLR due to increased runoff from deforestation
  W = SLR due to drainage of wetlands
  R = SLR reduction due to impoundment in reservoirs
  I = SLR reduction due to irrigation

The terms in Eq. (2), as used here, represent potential annual changes in SLR attributed to these various processes during the 1990s. These terms, strictly speaking, are not entirely independent. For example, much of mined groundwater is used in irrigation, a large fraction of which water is lost to evaporation and to seepage, as discussed below. A fuller treatment of these feedbacks and transient changes in atmospheric humidity and groundwater is beginning to be incorporated into hydrologic modeling studies, a few of which are briefly described below. Estimates for each of the terms in Eq. (2), based on historical data, are presented in the following two sections.

5.2 PROCESSES THAT INCREASE RIVER RUNOFF

5.2.1 Groundwater Mining

Groundwater mining or overdraft (G) is defined as the withdrawal or removal of groundwater in excess of natural recharge from infiltration and from underground inflow. Several parts of the world are already experiencing groundwater overdraft. This overexploitation of water resources is expected to increase as rising demand, spurred by population growth and development, encounters a diminishing supply, due to the increased evaporation, reduced precipitation, and less soil moisture projected for certain regions (e.g., the Mediterranean), from climate change (Houghton *et al.*, 1996). In the United States, major water withdrawals from the High Plains aquifer (formerly called the Ogallala aquifer) for irrigation have resulted in an average lowering of the regional water table by around 3.7 m since 1940 (Dugan *et al.*, 1994). This aquifer covers portions of South Dakota, Wyoming, Nebraska, Kansas, Colorado, Texas, Oklahoma, and New Mexico over an areal extent of 450,800 km^2. Southern California, Arizona, and New Mexico are other states where groundwater is mined (Table 5.3). Elsewhere, groundwater mining is a problem in arid countries of North Africa, the Sahel, and the Middle East (Table 5.3). Groundwater withdrawals are already very close to natural recharge rates in several other countries, such as Tunisia, Portugal, Lebanon, Syria, and Israel (WRI, 1998).

The total amount of water withdrawn (both surface and underground) and the fraction used consumptively (i.e., evaporated, transpired, or otherwise consumed by humans, animals, or plants, and not returned to streamflow) are documented in global water balance inventories (e.g., Shiklomanov, 1997; WRI 1998; Solley *et al.*, 1998, for the U.S.). The partitioning of water resources between surface and groundwater flow by continent is also known (Zekster and Loaiciga, 1993). Although groundwater mining data are lacking on a global level, a reasonable estimate can be made by extrapolating information on water usage, as follows: The volume of groundwater mined per year (V_{gm}) in selected countries or regions (Table 5.3, column 1) is listed in column 2. The total volume of water (both surface and underground) withdrawn annually $(V_{ww}$ in these countries or regions is given in column 3. The volume of groundwater withdrawn annually $(V_{gw,}$ col. 5) in each of these areas is estimated by multiplying the total volume of water withdrawn $(V_{ww,}$ col. 3) by the geographically corresponding fraction of the total water withdrawn that comes

Table 5.3

Estimated Contribution of Groundwater Mining to Sea Level Rise

Country/Region	V_{gm} (km³/yr)	V_{ww} (km³/yr)	V_{gw}/V_{ww}	V_{gw} (km³/yr)	V_{gm}/V_{gw}
High Plains	4.9				
SW U.S.A.	10.0				
California	13.0				
U.S.A.	27.9	471.7	0.224	105.7	0.26
Egypt	1.7	55.1	0.345	19.0	0.09
Libya	1.5	4.6	0.345	1.6	0.94
Mauritania	0.7	1.6	0.345	0.6	(1.1)
Algeria	0.3	4.5	0.345	1.6	0.19
Morocco	0.03	10.9	0.345	3.8	0.01
Africa	4.23	76.7	0.345	26.6	0.16
Greece	1.7	5.0	0.275	1.4	(1.2)
Italy	0.4	56.2	0.275	15.5	0.03
Spain	0.7	30.8	0.275	8.5	0.08
Europe	2.8	92.0	0.275	25.4	0.11
India	20.3	380.0	0.275	104.5	0.19
Saudi Arabia	5.2*	17.0	0.75*	12.8	0.41
Iran	0.2	70.0	0.275	19.3	0.01
Israel	0.2	2.0	0.275	0.6	0.33
Asia	25.9	469.0	0.275	137.2	0.19
Australia	0.04	14.6	0.149	2.2	0.02
TOTAL	60.9	1124	0.306	297.1	0.20

Data sources: Column 2: Dugan *et al.,* 1994; Shiklomanov, 1997; UN, 1983; Marinos and Diamandis, 1992: Beretta *et al.,* 1992; Singh, 1992; Jellali *et al.,* 1992; Lopen-Camacho *et al.,* 1992; Al-Ibrahim, 1991; Hosseinipour and Ghobackian, 1990, WRI, 1998.
Column 3: WRI, 1998; Solley *et al.,* 1998 (U.S.A.).
Column 4: Solley *et al.,* 1998 (U.S.A.); Zektser and Loaiciga, 1993 (elsewhere).
Column 5: col. 3 × col. 4; Column 6: col. 1 ÷ col. 5;
* from Al-Ibrahim, 1991.

from groundwater ($V_{gw}/V_{ww,}$ col. 4). The ratio (V_{gm}/V_{gw}) of the volume of groundwater mined (col. 1) to the volume of groundwater withdrawn (col. 5) in each region is then listed in column 6.

The global volume of groundwater withdrawn (V_{GW}) ranges between 991.4 km³/yr circa 1990 (3240 km³/yr [WRI, 1998] × 0.306 [the global V_{gw}/V_{ww} ratio, Zekster and Loaiciga, 1993]) and 1150.6 km³ in 1995 (3760 km³/yr × 0.306, after Shiklomanov, 1997). The global volume of groundwater mined (V_{GM}) can be extrapolated,

$$V_{GM} = V_{GW} \left(\Sigma V_{gm}/\Sigma V_{gw}\right), \tag{3}$$

where $\left(\Sigma V_{gm}/\Sigma V_{gw}\right)$ is the ratio of the sum of the volumes of groundwater mined in the countries or regions (col. 2) to the sum of the volumes of

Table 5.4

Anthropogenic Contributions to Sea Level Rise

	Equivalent sea level rise (mm/yr)		
Process	Low	Mid	High
Groundwater mining (G)	0.10	0.20	0.30
Urban runoff (U)	0.30	0.34	0.38
Water released by oxidation of fossil fuels, vegetation, and other sinks (C)	−0.06	0.010	0.07
Deforestation-induced runoff (D)	0.08	0.09	0.11
Wetlands loss (W)	0.001	0.0015	0.002
Reservoirs and dams (R)			
Storage	−0.33	−0.30	−0.27
Infiltration	−0.81	−0.68	−0.56
Evaporation	−0.010	−0.008	−0.005
Irrigation (I)			
Infiltration	−0.49	−0.44	−0.40
Evapotranspiration	−0.15	−0.12	−0.10
Total	−1.37	−0.91	−0.47

groundwater withdrawn in these areas (col. 5), or 0.20 (60.9/297.1). Thus, V_{GM} ranges between 198.3 km³ (0.20 × 991.4) and 230.1 km³ (0.20 × 1150.6), equivalent to 0.55 to 0.64 mm/yr SLR, respectively.[3]

Because country or regional ratios of V_{gm}/V_{gw} vary widely due to major differences in groundwater extraction practices, a more meaningful measure is the mean of the regional means of V_{gm}/V_{gw}, or 0.15 ± 0.09. For the stated range in V_{gm}/V_{gw} (Eq. (3), V_{GM} varies between 59.5 and 237.9 km³/yr (after WRI, 1998), which is equivalent to 0.17–0.66 mm/yr SLR, or 69.0 to 276.1 km³/yr (after Shiklomanov, 1997), equivalent to 0.19–0.77 mm/yr.

Around 61% of the water withdrawn globally in 1995 has been used consumptively (Shiklomanov, 1997), thereby leaving only 39% to run off. Therefore, the amount of mined groundwater that contributes to runoff, and ultimately to sea level, may only come to between 0.1 and 0.3 mm/yr to SLR, if all of this water were to flow into the ocean (see Table 5.4). However, in the future groundwater mining may well increase along with economic and agricultural development.

5.2.2   Urbanization

Although urban centers only occupy around 1% of the earth's total land area at present, they are home to a rapidly increasing proportion of the world's population (WRI, 1996). In spite of the relatively small land areas involved, urbanization (U) exerts a strong impact on hydrology as natural vegetation

[3] 360 km³ is equivalent to 1 mm of sea level rise.

is replaced with impermeable structures. Concrete and other artificial surfaces sharply reduce evapotranspiration and infiltration rates, contributing to falling water table levels, and also increases in surface runoff (Walling, 1987). Excessive groundwater extraction for municipal and industrial use has lowered water tables in many urban areas, such as Mexico City, Tucson, Houston, Bangkok, Venice, Shanghai, Tokyo, and Calcutta, causing land subsidence (and noteworthy for *coastal* cities, creating unusually high rates of relative sea level rise and salt water intrusion—see Chapter 8). By contrast, water tables have risen in other cities (e.g., London, Cairo, Riyadh, Saudi Arabia and other Persian Gulf cities, and Barcelona, Spain), due to reductions in pumping for public or industrial water supply, or to leakage of sewers or pipes (Chilton *et al.,* 1997). However, quantitative data on the global extent of groundwater changes accompanying urban water uses is lacking, except for specific case studies.

A significant consequence of urbanization is a net increase in total runoff due to the expansion of areas of impermeable pavements, buildings, and other covered surfaces, which reduce evapotranspiration and impede the infiltration of rainwater. Total urban runoff volumes may increase by factors of 2–2.5 (Jones, 1997, p. 226). The global increase in runoff associated with urban growth has been estimated to be around 137 km^3/yr (or 0.38 mm/yr SLR) in the 1980s (L'vovich and White, 1990). This figure probably represents an upper bound, in that it is based on relationships between changes in the extent of impermeable surfaces and runoff for several Russian cities, which may be higher than average for urban areas. Another complicating factor is that not all urban runoff necessarily feeds directly into rivers downstream. Overexploitation of groundwater in an aquifer hydraulically connected to the river will tend to reduce discharge downstream. The net effects of urban hydrologic changes are complex and difficult to predict on a worldwide basis. The uncertainty associated with projecting such changes suggests a plausible sea level change in the range of 0.30 to 0.38 mm/yr.

5.2.3 Water Released by Oxidation of Fossil Fuels and Vegetation

The combustion of fossils fuels and burning of tropical forests release water into the atmosphere together with carbon dioxide according to the general chemical equation

$$C_xH_y + (x + y/4)\ O_2 \rightarrow xCO_2 + y/2H_2O. \qquad (4)$$

Conversely, water and carbon dioxide are removed from the atmosphere as inorganic bicarbonate and carbonate ions in the ocean, and as new organic matter (reduced carbon) in forest regrowth (CO_2 assimilation) and soil carbon storage. Water exchanges associated with the carbon cycle (C) make a relatively minor contribution to sea level rise, as shown below (Table 5.5).

Table 5.5

Water Release Associated with CO_2 Emissions

	Average annual CO_2 budget (GTC/yr)	Water released, in SLR equivalent (mm/yr)
CO_2 sources		
Fossil fuel combustion and cement production	5.5 ± 0.5	0.021 (0.019–0.023)
Tropical deforestation	1.6 ± 1.0	0.033 (0.013–0.054)
Total sources	7.1 ± 1.1	
CO_2 sinks		
Atmosphere	3.3 ± 0.2	
Ocean uptake	2.0 ± 0.8	−0.008 (−0.012– −0.005)
Northern Hemisphere forest regrowth	0.5 ± 0.5	−0.010 (−0.021–0.0)
Additional sinks (CO_2 fertilization, nitrogen fertilization, soil storage)	1.3 ± 1.5	−0.026 (−0.058– −0.004)
Total sinks	7.1 ± 3.0	
Equivalent sea level rise	0.010 (−0.06 to +0.07) mm/yr	

Data from Houghton *et al.,* 1996.

5.2.3.1 Sources of CO_2 and H_2O

The three dominant fossil fuel types emit CO_2 in the following proportions: 0.16 (natural gas), 0.43 (liquid), and 0.41 (solid fuels) (after Marland and Boden, 1993), corresponding to emissions of 0.81, 2.28, and 2.14 GT C/yr, respectively (1 GT = 10^{15} g). These values can be used to estimate the volumes of water produced. (See Gornitz *et al.,* 1997, for details of calculation.)

The mass of CO_2 released by burning natural gas (methane) is 2.97 GT CO_2/yr, which is equivalent to 2.43 km^3/yr H_2O or 0.0068 mm/yr SLR. Combustion of petroleum yields 8.36 × GT CO_2/yr, which is equivalent to 3.63 km^3/yr H_2O or 0.01 mm/yr SLR. Burning coal generates 7.8 GT CO_2/yr, equivalent to 1.3 km^3/yr H_2O or 0.004 mm/yr SLR. The total water released by fossil fuel combustion is therefore equivalent to a SLR of 0.021 mm/yr (Table 5.5).

The net atmospheric carbon release from global land-use change is around 1.1 GT C/yr, representing the difference between tropical deforestation (1.6 GT C/yr) and temperate-boreal forest regrowth (0.5 GT C/yr, Table 5.5). This mass of carbon is equivalent to 1.65 km^3 H_2O (0.9–2.4 km^3 H_2O), or 0.005 (0.003–0.007 mm/yr SLR. Equal numbers of CO_2 and H_2O molecules are produced by the burning and/or bacterial decay of dry biomass.) The net annual water released by vegetation clearance each year, assuming a dry-to-wet biomass ratio of 0.25 (Rohrig, 1991), comes to 8.25 km^3 H$_2$O (4.5–12.0),[4] equivalent to 0.023 (0.013–0.033 mm/yr) SLR (Table 5.5).

5.2.3.2 Sinks of CO$_2$ and H$_2$O

The world's oceans absorb around 2.0 GT C/yr of CO$_2$ released by fossil fuel emissions and deforestation (Table 5.5). The oceanic uptake of CO$_2$ to form hydrogen ions and carbonate ions consumes the equivalent of 3.0 km^3/yr H$_2$O, which corresponds to about -0.008 mm/yr SLR.

Other terrestrial sinks for atmospheric carbon dioxide include enhanced plant assimilation under elevated CO$_2$ and enhanced plant growth due to nitrogen from fertilizers and fossil fuel burning. The uptake of 1.3 GT C (Table 5.5) corresponds to 1.95 km^3 of dry biomass. Using a dry-to-wet biomass ratio of 0.25 (Rohrig, 1991) gives the water taken up by new vegetation as 9.75 km^3 or -0.026 mm/yr SLR equivalent. The net contribution of water released by all of these oxidative processes is then around 0.01 mm/yr (in terms of SLR equivalent) (Table 5.4).

5.2.4   Changes in Runoff Due to Deforestation

Replacement of the natural vegetation cover by crops or grassland affects the fluxes of water and energy, as the leaf area, rooting depth, surface roughness, and albedo are altered. The vegetation changes associated with deforestation are generally accompanied by decreases in evapotranspiration and reduced soil infiltration. The infiltration capacity of a watershed may decrease substantially following deforestation, due to soil compaction by heavy logging or farm machinery, overgrazing and trampling by cattle, and increased soil erosion. The net effect of these changes is to increase runoff in newly cleared areas, at least in the short term, to create higher peak discharges in the wet season, leading to greater chances of flooding, and also to diminish streamflow during the dry season (Walling, 1987, pp. 55–57, 66–70; Jones, 1997, pp. 212–215). The long-term impacts on runoff are less clear, because of secondary regrowth and possible negative feedbacks on local rainfall, due to the reduced evapotranspiration and changes in the heat balance.

The extent of tropical forest has declined from around 1.80 billion hectares (ha) in 1990 to 1.73 billion ha in 1995 (WRI, 1998). This loss of tropical forest has been partially offset by forest regrowth in nontropical regions going from 1.71 billion ha in 1990 to 1.72 billion ha in 1995. Thus, the net forest cover reduction (primarily in the tropics) has been 56.4 million ha or 11.3 million ha/yr.

In the tropics, widespread forest clearing has been shown to increase total streamflow by as much as 125–820 mm/yr during the first three years (Bruijnzeel, 1996, 1993; Fritsch, 1993). The average increase in runoff is around

[4] This figure includes oxidation of above- and below-ground tree biomass, soil organic matter to 1 m depth, woody debris, and long-term decay of wood products.

300 mm/yr, when newly cleared humid tropical forest (D) is converted to cropland, tea, rubber, and cocoa plantations, or grassland (Bruijnzeel, 1996, 1993). Thus when this figure is multiplied by the net global average deforestation rate of 11.3×10^{10} m^2/yr between 1990 and 1995 (WRI, 1998), the runoff comes to approximately 33.8 km^3/yr, or 0.09 mm/yr SLR (Table 5.4). This estimate does not take into account vegetation regrowth or potential negative climate feedbacks that may decrease runoff, as suggested by several modeling studies (e.g., Henderson-Sellers *et al.*, 1993; Zhang *et al.*, 1996).

5.2.5 Wetlands

Wetland vegetation is adapted to the presence of a perennial near-surface water table, which creates standing bodies of water or saturated (waterlogged) soils. The location and areal coverage of wetlands are uncertain, due to the lack of a universally agreed upon definition in view of the wide diversity of wetland vegetation communities and soil types (Williams, 1990). Estimates of the global extent of wetland areas range from 5.26 million km^2 (Matthews and Fung, 1987) to 8.56 million km^2 (Williams, 1990).

The accumulation of organic-rich sediments in undisturbed wetlands constitutes a significant carbon sink. On the other hand, draining of swamps, bogs, or marshlands leads to oxidation of soil organic matter. Burning of vegetation, including peat for fuel, as well as agricultural conversion also releases CO_2 from wetlands.

Prior to large-scale conversion of forest clearance for agriculture (c. 1795), temperate wetlands accumulated carbon at rates of 56.9 to 82.6 million tons C/yr (Armentano and Menges, 1986). By 1980, annual carbon accumulation rates had decreased to -6.5 million tC/yr (i.e., a source of carbon) to 19.1 million tC/yr, representing a net reduction of 63.4 tC/yr in carbon storage from temperate wetlands (Armentano and Menges, 1986). Tropical clearance of wetlands, especially during the last several decades, brings the overall global reduction in annual carbon storage rates, relative to preclearance values, to around 150–184 million tC/yr (Armentano and Menges, 1986). If the ratio of annual carbon emission rate in temperate regions to the global rate (c. 1980) is assumed to be proportional to the temperate-to-global ratio of the difference between present carbon storage rates and preclearance rates, then the annual global emissions of carbon from wetland drainage and conversion could range between 22 and 101 million tC/yr.

Water released by drainage and burning of wetlands (computed like that of deforestation; see above) yields 0.03–0.15 km^3 H_2O per year, equivalent to 0.0001–0.0004 mm/yr SLR. With a dry-to-wet biomass ratio of 0.25, adding in the dry biomass brings the total SLR due to wetland drainage and oxidation to 0.001–0.002 mm/yr (Table 5.4). Possible changes in runoff resulting from wetland clearance have not been included, because of lack of relevant data.

5.3 PROCESSES THAT REDUCE RUNOFF

5.3.1 Reservoirs

Impoundment of water in reservoirs (R) constitutes a significant sequestration of freshwater on land that would otherwise flow to the world's oceans. (See Fig. 5.3, color plate.) Hence it represents a major reduction in potential SLR. The world's largest reservoirs (>1 million m³) provide a maximum storage capacity of 4821 km.³ The total comes to 5078 km³, if other types of dams are included (e.g., concrete dams >15 m and concrete face rockfill dams), which contribute an additional 39 and 218 km³ respectively (Intl. Water Power & Dam Constr. Yearbook, 1998). (The dam data base was checked to eliminate redundancies.) Shiklomanov (1997), on the other hand, estimates a maximum reservoir capacity of 6000 km³. Small impoundments (<100 million m³) could add another 4.5% (L'vovich and White, 1990, p. 239), bringing the total global reservoir storage capacity to between 5306 and 6270 km³. These figures represent upper bounds, inasmuch as most dams are not filled to capacity. A reasonable assumption is that reservoirs are filled to 85% of capacity, on average. Thus, the total global volume in reservoirs lies between 4510 and 5330 km³. Since more than 90% of the total reservoir capacity has been created since the 1950s, and since reservoir capacity has grown at a linear rate (Chao, 1995; Shiklomanov, 1997), the volume of water impounded in reservoirs could potentially reduce SLR by an average of 0.27 to 0.33 mm/yr, at present (Fig. 5.4, Table 5.4).

A major dam-building boom is under way, especially in the developing world. A conservative estimate of the additional reservoir volume to be created by dams under construction is around 953 km³, or 810 km³ if 85% full. This latter volume represents 15 to 18% of present reservoir storage. It is equivalent to a potential withholding of another 2.3 mm of SLR.

5.3.1.1 Evaporation from Reservoirs

Lakes, both natural and artificial, modify the climates of their surrounding areas, due to sharp land–water contrasts in temperature, albedo, and surface

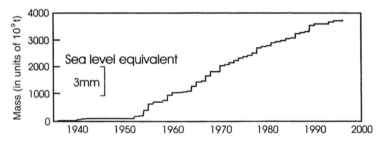

Figure 5.4 Cumulative mass of water held in reservoirs since 1950, in billions of metric tons and in SLR equivalent (after Chao, 1995).

roughness and also differences in the moisture budget of these two surface types. Evaporation over lakes or reservoirs is greater than that over land, particularly if the surrounding terrain is arid or if reservoir temperatures exceed those of the overlying air. Evaporation from reservoirs constitutes an additional continental store of water. Although the bulk of the evaporated water probably reprecipitates regionally, a small fraction may be held as vapor in the atmosphere, particularly in drier regions. This fraction may be reduced to some extent by long-range transport of atmospheric moisture to the sea.

Evaporative losses from reservoirs are calculated as the difference in mean evaporation between the lake surface and that of dry land, using mean evaporation data for each climate zone. These losses, amounting to around 6.5 km^3/ yr in 1950, increased to 164 km^3/yr in 1990 and 188 km^3/yr in 1995 (Shiklomanov, 1997). Although the fraction of the evaporated water that is stored in the atmosphere is uncertain, a plausible upper bound lies between 1 and 2% (see Section 5.3.2 on irrigation, below). This increased atmospheric moisture could reduce SLR by 0.005 to 0.010 mm/yr (Table 5.4).

5.3.1.2 Sedimentation Losses in Reservoirs

Rivers carry silt and other particulate matter in suspension. Normally, this material is deposited in natural basins, such as lakes, river deltas, and continental shelves. However, dams intercept these sediments, which then accumulate at the bottom of the man-made lakes. Sediment accumulation lowers the storage capacity of reservoirs, which in turn decreases the volume of water that otherwise would have been withheld from sea level rise. Although sedimentation rates vary widely, the overall effect is comparatively small. The extent of sediment settling depends on the sediment load of the river, the size and settling velocity of the sediments, and the trapping efficiency, which in turn is related to the reservoir capacity and annual volume of inflow.

Around 0.2% of U.S. reservoir storage capacity is lost annually to sedimentation (Gleick, 1992; Dendy et al., 1973). Lake Mead, for example, lost 0.3%/yr of capacity to siltation in the 1940s (Smith et al., 1960). Table 5.6 summarizes data on siltation rates from many reservoirs around the world. If the global mean annual loss of reservoir capacity due to siltation is no more than 1%, global reservoir storage would decrease by 45.1 to 53.3 km^3, or 1.13–1.33 km^3/yr since the 1950s, which corresponds to a SLR of 0.003–0.004 mm/yr. Dam construction may have decreased the sediment load of rivers delivered to the world's oceans by 885 million metric tons/yr (based on Milliman and Syvitski, 1992). Most of this sediment has been trapped behind dams. Given a mean rock density of 2.7 g/cm^3, this sediment diminishes the reservoir capacity by 0.33 km^3/yr, which is negligible in terms of its effect on SLR (0.001 mm/yr).

5.3.1.3 Seepage Losses in Reservoirs

Seepage losses also diminish the volume of water in reservoirs. The magnitude of these losses is related to the permeabilities of the enclosing bed rocks,

Table 5.6

Reduction in Reservoir Capacity due to Siltation

Reservoir	Period	Annual reduction in capacity (%/yr)
High Aswan Dam, Nile River, Egypt	1967–1991	0.05
Sampling of U.S. reservoirs	19.7 yr period	0.2
Lake Mead, Colorado River	1935–1949	0.3
Bhakra Lake, Sutlej River, India	1959–1969	0.5
Sennar (Makwar Reservoir), Blue Nile, Sudan	1925–1986	0.6
Imagi Reservoir, Kenya	—	0.8
Roseires Reservoir, Blue Nile, Sudan	1966–1981	1.3
Bhumipol Reservoir, Thailand	—	1.5
Tarbela Dam, Indus River, Pakistan	1974–1990	2.0
Khashm-el-Girba Reservoir, Atbara River, Sudan	1965–1990	2.2
Sanmexia Dam, Yellow River, China	1960–1990	2.2
Nizam Sagar Dam, Manjira, India	1931–1961	2.2
Shah-Banou Farah Reservoir, Iran	1961–1990	2.4
Estimated global average rate		1%

Data sources: Shahin, 1993; Rao and Palta, 1973; Dendy *et al.*, 1973; Smith *et al.*, 1960; Douglas, 1990; James and Kiersch, 1988.

geologic structures, such as fractures or faults, and the height of the water table (James and Kiersch, 1988). Seepage losses occur where solutions can migrate away from the sides or base of the reservoir. These occur along fractures, faults, solution cavities or sinkholes in carbonate rocks (Money-maker, 1969), unconsolidated and/or permeable sedimentary rocks (Gardner, 1969), and lava tunnels (Monahan, 1969).

Very few data exist on seepage losses. Lake Nasser may be losing up to ~0.6% of capacity per year (Wafa and Labib, 1973). However, storage of water in dry rock pore volume may ultimately reach up to 29% of maximum reservoir capacity at Lake Nasser. Lake Mead lost ~1.7% of its initial capacity per year in the 1940s (Smith *et al.*, 1960). Annual seepage losses for several small mid-Western reservoirs range from 0.1 to 39% of capacity (Gardner, 1969). Unfavorable geologic conditions leading to severe reservoir leakage have resulted in reservoir abandonment and even dam failure, in several extreme cases (Kiersch, 1958).

Although seepage data are fragmentary, an annual average loss has been estimated at up to ~5% of reservoir volume (Gleick, 1992). In the absence of additional data, we assume an average annual loss of 5 ± 0.5% of reservoir volume for reservoirs built up to the 1990s. This amounts to 203–293 km^3/yr, or around 0.56–0.81 mm/yr withheld from SLR (Table 5.4).

Percolation beneath reservoirs recharges aquifers, which ultimately increases the groundwater discharge to the sea. However, because the average global residence time for groundwater in aquifers has been estimated at around

330 years (L'vovich, 1979) and possibly even thousands of years or more (Jones, 1997), this effect is probably relatively small over the short (<100 year) periods considered here and has therefore not been taken into account. Nevertheless, some of this seepage may actually enter groundwater discharge and/or river runoff, even within a few years, thereby not adding to storage of water on land. Thus the estimated water sequestration from this source could be high.

5.3.2   Irrigation

Irrigation is a major consumer of worldwide freshwater resources, accounting for around 39% of all freshwater use in the United States (Solley *et al.*, 1998) and 69% internationally (WRI, 1998). As in the case of reservoirs, losses of irrigation water (I) to deep seepage and evaporation also lessen the anthropogenic input to SLR. In 1996, 263 million ha were under irrigation, with 70% of the total in Asia, 9.5% in Europe, 5% in Africa, and 15% in North, Central, and South America (FAO, 1997).[5] The average annual use of water in irrigation varies widely, ranging from around 12,000 to 15,000 m^3/ha in North Africa to less than 4000 m^3/ha in several more water-efficient western European countries (Framji *et al.*, 1981). The area-weighted global average water use in irrigation is approximately 9450 m^3/ha. This figure multiplied by the global area under irrigation gives a volume of 2484 km^3. Conveyance losses in open canals and ditches may increase water losses by a factor of 1.3 (Postel, 1989). Thus, the total comes to 3229 km^3.

A substantial fraction of the water applied in irrigation is used consumptively, that is, it is taken up and evapotranspired by crops, and thus does not return to streamflow (at least not in the short run; a small, but not well-known quantity could enter via underground or surface discharge). Globally, the consumptive use of water in irrigation is around 78 ±6% of the water withdrawn (Shiklomanov, 1997). It varies regionally between 61% in North America and 82% in Asia. Using an average value of 78% combined with the above estimate of water used in irrigation gives 2519 km^3/yr, with a range between 2325 and 2712 km^3/yr. A small, as yet unquantified amount of water from evapotranspiration remains in the atmosphere. An accurate assessment of this fraction is needed; if we assume that up to 2% of the evaporated water remains in the atmosphere as added vapor, then the amount of SLR withheld through evapotranspiration would be as much as 0.12 mm/yr (0.10–0.15 mm/yr) in the 1990s (Table 5.4).

Irrigation water not lost in evapotranspiration (16–28%) infiltrates into the soil (equivalent to 516.6–904.1 km^3). If around 5 ±0.5% of this water is assumed to percolate to depth, as in the case of reservoirs, this yields 23–50 km^3. In addition up to 98% of the irrigation water lost through evapotranspiration eventually reprecipitates. If again 5 ±0.5% of this recycled water infiltrates

[5] The 1996 figure for the United States is missing, so the 1995 value was added to the global total.

at depth, this removes another 103–146 km^3. A total of 143–175 km^3 could therefore be sequestered at depth, corresponding to 0.40–0.49 mm/yr withheld from SLR (Table 5.4). As indicated above, because of seepage, these estimates are upper bounds.

5.4 DISCUSSION

The effects on sea level caused by the anthropogenic modifications of the hydrologic cycle are summarized in Table 5.4. Groundwater mining, increased runoff from land-cover changes, and to a lesser extent, burning of fossil fuel and vegetation could raise SLR by 0.6 ±0.2 mm/yr. Conversely, impoundment of water in reservoirs and losses due to infiltration and evapotranspiration from artificial lakes and irrigation could keep the equivalent of 1.5 ±0.2 mm/yr from reaching the ocean. The combined effect of all of these processes could withhold around 0.9 ±0.5 mm/yr from reaching the ocean. These results imply a reduction in sea level rise comparable in magnitude to the observed recent sea level rise, but of opposite sign.

Figure 5.5 illustrates the evolution of estimated historical changes in sea level resulting from various anthropogenic activities since 1900, based on the midrange estimates of Table 5.4. The effects on sea level become significant only within the last 50 years. The growth in anthropogenic impacts is closely linked to the overall human transformation of the earth's surface, driven in general by the rapid 20th-century expansion in population and economic development (Turner *et al.,* 1990). When the time inevitably arrives that reservoir construction significantly slows, there is a great likelihood of an important increase in the rate of global sea level rise.

The calculations outlined above contain large uncertainties, because of the fragmentary and incomplete nature of the historical data sets. The full extent of groundwater mining may have been underestimated due to limited country information. On the other hand, overestimates in runoff due to deforestation and to urbanization, which tend to raise sea level, may at least partially offset overestimates in rates of atmospheric moisture buildup and losses to deep percolation, yielding a net result roughly comparable to that suggested here.

Several approaches can be taken to reduce uncertainties in the estimates of human-induced changes in terrestrial water storage. A more through investigative search of historical records could provide additional information on groundwater mining and seepage losses beneath reservoirs and in irrigation. However, a unified systems approach using computer models will be needed to trace the path of the water more accurately through the atmosphere, hydrologic, and biosphere subsystems and to account for various feedbacks. A start in this direction is provided by several recent general circulation model (GCM) experiments which examine the effects of land cover changes, particularly tropical deforestation, on the hydrologic cycle and on regional to global climate

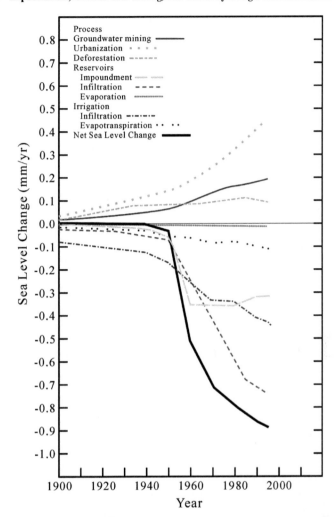

Figure 5.5   Evolution of potential contributions to sea level rise from various anthropogenic processes, using midrange estimates of Table 5.4. Historical data on urban runoff and reservoirs, after Shiklomanov, 1997; global groundwater withdrawn, irrigation water consumption, after FAO (1997); deforestation (WRI, 1998). See text for further details.

(Zhang *et al.*, 1996; Lean *et al.*, 1996; Polcher and Laval, 1994; Henderson-Sellers *et al.*, 1993). More advanced surface and groundwater hydrology models embedded within GCMs could begin to analyze the movement of water among the various surface and subsurface reservoirs.

Finally, satellite remote sensing offers a potentially useful technology for monitoring the global hydrologic budget (Koster *et al.*, 1999). Sampling of terrain by TOPEX/POSEIDON, Landsat, SPOT, and other land-surface imaging satellites could provide better estimates of smaller artificial lakes and impoundments, omitted in the current calculations, and also changes in deforestation and other land-use transformations (Chen *et al.,* 1998; Minster *et al.,* 1999).

5.4.1   Modeling Transformations of the Land Hydrological Cycle

In many climate model simulations, surface temperatures increase, whereas precipitation and evapotranspiration decrease in the areas affected by deforestation (Zhang *et al.,* 1996; Lean *et al.,* 1996; Henderson-Sellers *et al.,* 1993). Evapotranspiration diminishes because of reduced surface roughness and an increase in albedo, as rainforest is altered to pasture. These changes contribute to the calculated decrease in rainfall following deforestation (Lean *et al.,* 1996). In Amazonia, model results and ground observations suggest that nearly two-thirds of the precipitation is recycled via rainforest evapotranspiration and soil evaporation, the balance going into surface and groundwater runoff (Zhang *et al.,* 1996; Salati and Nobre, 1991).

Generally, the reduction in calculated annual precipitation exceeds the reduction in evapotranspiration, leading to decreases in runoff after forest clearance (see summary reviews in Henderson-Sellers *et al.,* 1993; Lean *et al.,* 1996). (One study found substantial increases in runoff, resulting from larger decreases in evapotranspiration than in precipitation; Lean *et al.,* 1996.) These model results, however, generally run counter to field observations, which record increases in runoff (e.g., Bruijnzeel, 1996, 1993). Changes in soil structure, such as loss of topsoil organic matter, compaction due to overgrazing and mechanical disturbances, and soil erosion, would combine to sharply reduce soil infiltration rates, thus increasing surface runoff, especially during wet seasons, and decreasing soil moisture in the dry seasons.

"Tracer" techniques employ climate models and variations in trace isotope concentrations to track sources and movement of atmospheric moisture (Salati *et al.,* 1979; Koster *et al.,* 1986; Druyan and Koster, 1989). These techniques, in conjuction with improved hydrologic models, could eventually help address the specific questions posed here—to what extent do large-scale anthropogenic transformations of the hydrologic cycle, such as impoundment of water in large reservoirs, irrigation, or deforestation alter the net balance between evaporation, precipitation, and/or advection of moisture to adjoining regions? Changes in this balance will determine (1) whether dam-building or irrigation would increase atmospheric moisture storage over land (especially in arid regions), which would reduce water transport to the ocean, and thus limit sea level rise, or (2) whether deforestation would decrease local evaporation/ evapotranspiration and precipitation, increase runoff, and hence augment rates of sea level rise. More advanced surface and groundwater hydrology models embedded within GCMs may be able to analyze additional feedbacks, such as the partitioning of water from groundwater mining among runoff, infiltration, evaporation, and the amount of water infiltrated beneath reservoirs and irrigated fields that contributes to underground flow.

5.4.2   Hydrologic Monitoring by Satellites

A new and potentially useful methodology involves the application of satellite altimetry and gravity data to monitoring the global water mass budget (Chen *et al.*, 1998; Minister *et al.*, 1999; Herring, 1998). Satellite radar altimeters, such as TOPEX/POSEIDON, can measure the ocean surface height to millimetric accuracy (Nerem, 1999). The global average sea level rise determined from TOPEX/POSEDON for 1993–1998 lies between 2.5 and 3.1 mm/yr, which is slightly higher than the range reported by Houghton *et al.* (1996), based on tide gauges. The TOPEX/POSEIDON satellite altimeter can also derive the ocean component of the global hydrologic cycle and help validate the global water mass budget (Minister *et al.*, 1999; Chen *et al.*, 1998). After the altimeter data are corrected for various orbital and atmospheric effects, the steric sea level variation, caused by temperature and salinity differences, is removed using historical *in situ* water density measurements. The residual sea level represents the true change in global ocean water mass. This value can be employed to constrain the contributions from the atmospheric and continental components of the global water budget, which are derived from various climatological data bases. The land component, dependent upon differences between precipitation, evaporation, and stream runoff, is the least well known, as demonstrated above. Future improved satellite observations will supplement ground-based measurements and hydrologic models, in order to detect anthropogenic changes, superimposed on natural climate variations.

Two satellites, to be launched within the next few years, will measure annual variations in the earth's gravity due to changes in mass as small as a layer of 1 cm of water over 250,000 km^2 (Herring, 1998). Changes in gravity due to variations in sea level and in continental water table levels, such as the lowering of the High Plains aquifer of the Great Plains, which has dropped by 30 m in some places over the last 40 years (Dugan *et al.*, 1994), should be detectable by making comparisons to earlier gravity surveys.

5.5   CONCLUSIONS

Plausible estimates of sea level changes accompanying large-scale anthropogenic modifications of land hydrologic processes have been presented. Increased runoff from groundwater mining and impermeable urbanized surfaces are potentially important human-induced sources of sea level rise. Runoff from tropical deforestation, wetlands clearance, and water released by oxidation of fossil fuel and biomass, are relatively insignificant. Altogether, these processes could augment sea level by some 0.4 to 0.9 mm/yr (Table 5.4). Conversely, sequestration of water by dams, and further losses of water due to infiltration beneath reservoirs and irrigated fields, along with evaporation from these surfaces could retain the equivalent of 1.3 to 1.8 mm/yr over continents. The net effect of all of these anthropogenic processes is to withhold the equivalent of 0.9 ±0.5 mm/yr from the sea. This rate represents a substantial fraction of the observed recent sea level rise of 1 to 2.5 mm/yr, but opposite in sign.

However, it should be pointed out that these estimated impacts on sea level rise represent upper bounds. The historical data base is incomplete and subject to considerable uncertainties. Furthermore, the increased volume of moisture stored in the atmosphere due to evaporation from reservoirs or irrigated fields provides only an upper bound, inasmuch as the atmospheric effects are probably localized and thus are unlikely to represent large-scale averages. Finally, a number of the processes are interrelated, requiring a unified systems approach, using coupled land–surface hydrology and climate models, to track the path of water through the various subsystems, accounting for possible feedbacks.

Acknowledgments

This research was supported by NASA-Columbia Cooperative Agreement NCC 5-328.

REFERENCES

Al-Ibrahim, A. A. (1991). Excessive use of groundwater resources in Saudi Arabia: Impacts and policy options. Ambio **20**, 34–37.

Armentano, T. V. and Menges, E. S. (1986). Patterns of change in the carbon balance of organic soil-wetlands of the temperate zone. *J. Ecol.* **74**, 755–774.

Beretta, G. P., Pagotto, A., Vandini, R, and Zanni, S. (1992). Aquifer overexploitation in the Po Plain: Hydrogeological, geotechnical, and hydrochemical aspects. In I. Simmers, F. Villarroya, and L. F. Rebollo (eds.), *Selected Papers on Aquifer Overexploitation,* 23rd Intl. Congress of I.A.H., Puerto de la Cruz, Tenerife, Spain, April 15–19, 1991. *Hydrogeology, Selected Papers* **3**, 115–130.

Bruijnzeel, L. A. (1993). Land-use and hydrology in warm humid regions: Where do we stand? In J. S. Gladwell (ed.), *Hydrology of Warm Humid Regions,* Intl Assoc. Hydrol. Sci. Publ. No. 216, 3–34.

Bruijnzeel, L. A. (1996). Predicting the hydrological impacts of land cover transformations in the humid tropics: The need for integrated research. In J. H. C. Gash, C. A. Nobre, J. M. Roberts, and R. L. Victoria (eds.), *Amazonian Deforestation and Climate,* pp. 15–55. Wiley, Chichester.

Chao, B. F. (1991). Man, water, and global sea level. *EOS* **72**, 472.

Chao, B. F. (1994). Man-made lakes and sea-level rise. *Nature* **370**, 258.

Chao, B. F. (1995). Anthropogenic impact on global geodynamics due to reservoir water impoundment. *Geophys. Res. Lett.* **22**, 3529–3532.

Chen, J. L., Wilson, C. R., Chambers, D. P., Nerem, R. S., and Tapley, B. D. (1998). Seasonal global water mass budget and mean sea level variations. *Geophys. Res. Lett.* **25**, 3555–3558.

Chilton, J., *et al.* (eds.) (1997). *Groundwater in the Urban Environment,* Vol. 1, *Problems, Processes, and Management.* A. A. Balkema, Rotterdam.

Dendy, F. E., Champion, W. A., and Wilson, R. B. (1973). Reservoir sedimentation surveys in the United States. In W. C. Ackermann, G. F. White, and E. B. Worthington (eds.), *Man-Made Lakes: Their Problems and Environmental Effects.* Geophys. Monograph **17**, 349–356.

Douglas, I. (1990). Sediment transfer and siltation. In B. L. Turner II, W. C. Clark, R. W. Kates, J. F. Richards, J. T. Mathews, and W. B. Meyer (eds.), *The Earth as Transformed by Human Action.* pp. 215–234, Cambridge Univ. Press and Clark University, Cambridge.

Druyan, L. M. and Koster, R. D. (1989). Sources of Sahel precipitation for simulated drought and rainy seasons. *J. Clim.* **2**, 1438–1446.

Dugan, J. T., McGrath, T., and Zelt, R. B. (1994). Water-level changes in the High Plains Aquifer—Predevelopment to 1992. U.S. Geol. Surv. Water-Resource Investigations Report 94-4027.

Food and Agriculture Organization (1997). *FAO Production Yearbook,* Vol. 51, FAO, UN, Rome.

Framji, K. K., Garg, B. C., and Luthra, S. D. L. (1981), *Irrigation and Drainage in the World—A Global Review,* 3rd ed., Vol. 1, Introduction, pp. xli–cxxv. Intl. Com. Irrig. Drainage.

Fritsch, J. M. (1993). The hydrological effects of clearing tropical rainforest and of the implementation of alternative land uses. In J. S. Gladwell (ed.), *Hydrology of Warm Humid Regions.* Intl. Assoc. Hydrol. Sci. Publ. No. 216, 53–66.

Gardner, W. I. (1969). Dams and reservoirs in Pleistocene eolian deposit terrain of Nebraska and Kansas. *Bull. Assoc. Engin. Geol.* **6**, 31–44.

Gleick, P. H. (1992). Environmental consequences of hydroelectric development: The role of facility size and type. *Energy* **17**, 735-747.

Gleick, P. H. (ed.) (1993). *Water in Crisis.* Oxford Univ. Press, New York.

Gornitz, V. (1995). Sea-level rise: A review of recent past and near-future trends. *Earth Surf. Proc. Landforms* **20**, 7–20.

Gornitz, V. and Lebedeff, S. (1987). Global sea level changes during the past century. In D. Nummedal, O. H. Pilkey, and J. D. Howard (eds.), *Sea Level Fluctuation and Coastal Evolution,* pp. 3–16. SEPM Spec. Publ. 41.

Gornitz, V., Rosenzweig, C., and Hillel, D. (1997). Effects of anthropogenic intervention in the land hydrologic cycle on global sea level rise. *Glob. Planet. Change* **14**, 147–161.

Henderson-Sellers, A., Dickinson, R. E., Durbidge, T. B., Kennedy, P. J., McGuffie, K., and Pitman, A. J. (1993). Tropical deforestation: Modelling local- to regional-scale climate change. *J. Geophys. Res.* **98**, 7289–7315.

Herring, T. A. (1998). Appreciate the gravity. *Nature* **391**, 434–435.

Hosseinipour, Z. and Ghobadian, A. (1990). Groundwater depletion and salinity in Yazd, Iran. In *Hydraulics/Hydrology of Arid Lands, Proc. Intl. Symp.,* pp. 465–471. Am. Soc. Civ. Eng., New York.

Houghton, J. J., Meira Filho, L. G., Callander, B. A., Harris, N., Kattenberg, A., and Maskell, K. (eds.) 1996. *Climate Change 1995—The Science of Climate Change,* pp. 359–405. Cambridge Univ. Press, Cambridge.

International Water Power and Dam Construction Yearbook (1998). *The World's Major Dams and Hydro Plants.* Wilmington Business Publishing, Kent, U.K.

James, L. B. and Kiersch, G. A. (1988). Reservoirs: Leakage from reservoirs. In J. B. Jansen (ed.), *Advanced Dam Engineering for Design, Construction, and Rehabilitation,* pp. 722–729. Van Nostrand Reinhold, New York.

Jellali, M., Geanah, M., and Bichara, S. (1992). Groundwater mining and development in the Souss valley (Morocco). In I. Simmers, F. Villarroya, and L. F. Rebollo (eds.), *Selected Papers on Aquifer Overexploitation* 23rd Intl. Congress of I.A.H., Puerto de la Cruz, Tenerife, Spain, April 15–19, 1991. *Hydrogeology,* Selected Papers **3** 337–348.

Jones, J. A. A. (1997). *Global Hydrology: Processes, Resources and Environmental Management.* Addison Wesley Longman, Harlow, Essex, U.K.

Kiersch, G. A. (1958). Geologic causes for failure of Lone Pine Reservoir, east-central Arizona. *Econ. Geol.* **53,** 854–866.

Koster, R. D., Houser, P. R., Engman, E. T. and Kustas, W. P. (1999). Remote sensing may provide unprecedented hydrological data. *EOS,* **80,** 156.

Koster, R., Jouzel, J., Suozzo, R., Russell, G., Broecker, W., Rind, D., and Eagleson, P. (1986). Global sources of local precipitaiton as determined by the NASA/GISS GCM. *Geophys. Res. Lett.* **13,** 121–124.

Lean, J., Bunton, C. B., Nobre, C. A., and Rowntree, P. R. (1996). The simulated impact of Amazonian deforestation on climate using measured ABRACOS vegetation characteristics. In J. H. C. Gash, C. A. Nobre, J. M. Roberts, and R. L. Victoria (eds.), *Amazonian Deforestation and Climate,* pp. 549–576. Wiley, Chichester.

Lopen-Camacho and Sanchez-Gonzalez, A. (1992). Overexploitation problems in Spain. In I. Simmers, F. Villarroya, and L. F. Rebollo (eds.), *Selected Papers on Aquifer Overexploitation,* 23rd Intl. Congress of I.A.H., Puerto de la Cruz, Tenerife, Spain, April 15–19, 1991. *Hydrogeology,* Selected Papers **3,** 363–371.

L'vovich, M. I. (1979). *World Water Resources and Their Future.* Am. Geophys. Union. (tr. R. L. Nace).

L'vovich, M. I. and White, G. F. (1990). Use and transformation of terrestrial water systems. In B. L. Turner II, W. C. Clark, R. W. Kates, J. F. Richards, J. T. Mathews, and W. B. Meyer (eds.), *The Earth as Transformed by Human Action.* Cambridge Univ. Press and Clark University, pp. 235–252. Cambridge.

Marengo, J. A., Miller, J. A., Russell, G. L., Rosenzweig, C. E., and Abramopoulos, F. (1994). Calculations of river-runoff in the GISS GCM: Impact of a new land-surface parameterization and runoff routing model on the hydrology of the Amazon River. *Clim. Dynamics* **10,** 349–361.

Marinos, P. and Diamandis, J. (1992). The risk of overexploitation of a multi-layer heterogeneous aquifer system: Hydrological considerations in major Greek basins. In I. Simmers, F. Villarroya, and L. F. Rebollo (eds.), *Selected Papers on Aquifer Overexploitation,* 23rd Intl. Congress of I.A.H., Puerto de la Cruz, Tenerife, Spain, April 15–19, 1991. *Hydrogeology,* Selected Papers **3,** 61–66.

Marland, G. and Boden, T. (1993). The magnitude and distribution of fossil-fuel related carbon releases. In M. Heimann (ed.), *The Global Carbon Cycle,* pp. 117–138. Springer-Verlag, Berlin.

Matthews, E. and Fung, I. (1987). Methane emission from natural wetlands: Global distribution, area, and environmental characteristics of sources. *Glob. Biogeochem. Cycles* **1,** 61–86.

Milliman, J. D., and Syvitski, J. P. M. (1992). Geomorphic/tectonic control of sediment discharge to the ocean: The importance of small mountain rivers. *J. Geol.* **100,** 525–544.

Minster, J. F., Cazenave, A., Serafini, Y. V., Mercier, F., Gennero, M. C., and Rogel, P. (1999). Annual cycle in mean sea level from Topex-Poseidon and ERS-1: Inference on the global hydrological cycle. *Glob. Planet. Change* **20,** 57–66.

Monahan, C. J. (1969). Reservoirs in volcanic terrain. *Bull. Assoc. Eng. Geol.* **6,** 53–69.

Moneymaker, B. C. (1969). Reservoir leakage in limestone terrains. *Bull. Assoc. Eng. Geol.* **6,** 3–30.

Nerem, R. S. (1999). Measuring very low frequency sea level variations using satellite altimeter data. *Glob. Planet. Change* **20,** 157–171.

Newman, W. S. and Fairbridge, R. W. (1986). The management of sea-level rise. *Nature* **320,** 319–321.

Newson, M. (1992). *Land, Water, and Development: River Basins and their Sustainable Development.* Routledge, London.

Peltier, W. R. and Tushingham, A. M. (1991). Influence of glacial isostatic adjustments on tide-gauge measurements of secular sea level change. *J. Geophys. Res.* **96,** 6779–6796.

Polcher, J. and Laval, K. (1994). The impact of African and Amazonian deforestation on tropical climate. *J. Hydrol.* **155,** 389–405.

Postel, S. (1989). *Water for Agriculture: Facing the Limits.* Worldwatch Paper 93, Worldwatch Institute, Washington, DC.

Postel, S. L., Daily, G. C., and Ehrlich, P. R. (1996). Human appropriation of renewable freshwater. *Science,* **27,** 785–788.

Rao, K. L. and Palta, B. R. (1973). Great man-made lake of Bhakra, India. In W. C. Akermann, G. F. White, and E. B. Worthington, (eds.), *Man-Made Lakes: Their Problems and Environmental Effects.* Geophys. Monograph **17,** 170–185.

Rodenburg, E. (1994). Man-made lakes and sea-level rise. *Nature* **370,** 258.

Rohrig, E. (1991). In E. Ulrich (ed.), *Ecosystems of the World,* pp. 165–174. Elsevier, New York.

Sahagian, D. L., Schwartz, F. W., and Jacobs, D. K. (1994). Direct anthropogenic contributions to sea level rise in the twentieth century. *Nature* **367,** 54–57.

Salati, E., Dall'Ollio, A., Matsui, E. and Gat, J. R. (1979). Recycling of water in the Amazon Basin: An isotopic study. *Water Resour. Res.* **15,** 1250–1258.

Salati, E. and Nobre, C. A. (1991). Possible climatic impacts of tropical deforestation. *Clim. Change* **19,** 177–196.

Schumann, A. H. (1998). Land-use change. In R. W. Herschy and R. W. Fairbridge (eds.), *Encyclopedia of Hydrology and Water Resources,* pp. 466–467. Kluwer Academic, Dordrecht.

Shahin, M. M. A. (1993). An overview of reservoir sedimentation in some African river basins. In R. F. Haldey and T. Mizuyama (eds.), *Sediment Problems: Strategies for Monitoring, Prediction, and Control.* IAHS Publ. No. 217, 93–100.

Smith, W. O., Vetter, C. P., Cummings, G. B., *et al.* (1960). Comprehensive survey of sedimentation in Lake Mead, 1948–49. U.S. Geol. Survey Prof. Paper 295.

Shiklomanov, I. A. (1993). World fresh water resources. In P. H. Gleick (ed.), *Water in Crisis.* Oxford Univ. Press, New York.

Shiklomanov, I. A. (1997). *Comprehensive Assessment of the Freshwater Resources of the World: Assessment of Water Resources and Water Availability in the World.* U.N. World Meteorological Organization.

Singh, K. P. (1992). Groundwater overdraft in northwest parts of Indo-gangetic alluvial plains, India: Feasibility of artificial recharge. In I. Simmers, F. Villarroya, and L. F. Rebollo (eds.), *Selected Papers on Aquifer Overexploitation,* 23rd Intl. Congress of I.A.H., Puerto de la Cruz, Tenerife, Spain, April 15–19, 1991. *Hydrogeology,* Selected Papers **3,** 83–89.

Solley, W. B., Pierce, R. R., and Perlman, H. A. (1998). *Estimated Use of Water in the United States in 1995.* U.S. Geol. Surv. Circular 1200.

Turner II, B. L., Clark, W. C., Kates, R. W., Richards, J. F., Mathews, J. T., and W. B. Meyer (eds.) (1990), *The Earth as Transformed by Human Action: Global and Regional Changes in the Biosphere over the Past 300 Years.* Cambridge Univ. Press and Clark University, Cambridge.

United Nations (1983). *Ground Water in the Pacific Region,* U.N. Natural Resources/Water Series No. 12, United Nations, New York.

Wafa, T. A. and Labib, A. H. (1973). Seepage losses from Lake Nasser. In W. C. Ackermann, G. F. White, and E. B. Worthington (eds.), *Man-Made Lakes: Their Problems and Environmental Effects.* Geophys. Monograph **17,** 287–291.

Walling, D. E. (1987). In K. J. Gregory and D. E. Walling (eds.), *Human Activity and Environmental Processes,* pp. 53–85. Wiley, Chichester, New York.

Warrick, R. A., Le Provost, C., Meier, M. F., Oerlemans, J., and Woodworth, P. L. (1996). Changes in Sea Level. In Houghton, J. J., Meira Filho, L. G., Callander, B. A., Harris, N., Kattenberg, A., and Maskell, K. (eds.), *Climate Change 1995—The Science of Climate Change,* pp. 359–405. Cambridge Univ. Press, Cambridge.

Williams, M. (ed.) 1990. *Wetlands—A Threatened Resource,* Basil Blackwell, Oxford.

World Resources Institute (1998). *World Resources 1998–1999,* Table 12.1, pp. 304–305; Table 12.2, pp. 306–307. Oxford Univ. Press, New York.

Zektser, I. S. and Loaiciga, H. A. (1993). Groundwater fluxes in the global hydrologic cycle: Past, present and future. *J. Hydrol.* **144,** 405–427.

Zhang, H., and Henderson-Sellers, A. (1996). Impacts of tropical deforestation. I. Process analysis of local climatic change. *J. Clim.* **9,** 1497–1517.

Chapter 6 | Observation of Sea Level Change from Satellite Altimetry

R. S. Nerem
G. T. Mitchum

6.1 INTRODUCTION

6.1.1 Tide Gauges and Satellite Altimetry

Long-term sea level change has ordinarily been estimated from tide gauge data. However, two fundamental problems are encountered using tide gauge measurements for this purpose. First, tide gauges only measure sea level change relative to a crustal reference point, which may be moving vertically at rates comparable to the true sea level signals (Douglas, 1995). Second, tide gauges have limited spatial distribution and suboptimal coastal locations (Barnett, 1984; Groger and Plag, 1993) and thus provide poor spatial sampling of the open ocean. Douglas (1991, 1992) has argued that by selecting tide gauge records of at least 50 years in length and away from tectonically active areas, even a limited set of poorly distributed tide gauges can give a useful estimate of global sea level rise. However, averaging over such a long time period makes the investigation of shorter term changes problematic at best.

Over the past decade researchers interested in sea level variations have had an exciting new tool added to their arsenal. During this time satellite altimetry—the measurement of sea surface height from space—has come of age as an oceanographic observing technology. Beginning in 1992, with the launch of a joint U.S. and French project called TOPEX/POSEIDON (hereinafter, T/P), researchers have had routine access to high-quality data proven useful for many applications. For readers interested in the broad scope of the applications of the T/P data, a good starting point for further reading is the special issue of the *Journal of Geophysical Research* (1995, Vol. 100, No. C12), which was dedicated to science results and investigations using T/P data. In this chapter we will focus on one particular application, using T/P data for measurement of long-term sea level change. We will primarily focus on low-frequency changes in sea level associated with changes in total ocean volume, as discussed by Douglas in Chapter 3, rather than on interannual to decadal changes associated with dynamical changes in the large-scale ocean circulation like those discussed by Sturges and Hong in Chapter 7. We will, however, briefly mention some interesting interannual changes we have observed in total ocean volume associated with the 1997–1998 El Niño/Southern Oscillation (ENSO) event.

Why is satellite altimetry such an interesting new tool for the study of long-term sea level change? To answer this question let us consider some issues related to how altimeters, as opposed to tide gauges, sample the global sea surface height field.

Begin by imagining a densely spaced, global set of perfect tide gauges, that is, tide gauges all geodetically leveled together and perfectly measuring the sea surface height variations typical of the open ocean. Then aside from lacking perfection, the actual tide gauge network is simply a subset of this hypothetical dense global set. Now imagine that the total volume of the real ocean did not change at all in time, but that the ocean mass was allowed to redistribute. For example, there could be a tilt from east to west across an ocean basin that increased slowly in time. In this case the average sea level computed from the dense set of gauges would properly return a value that did not vary in time. The real tide gauge network, on the other hand, could not do so. Depending on how the actual gauges were distributed with respect to the east–west tilt, there would be a trend in the average sea level that could be mistakenly interpreted as ocean volume change. The main purpose of this very simple exercise is to point out that because satellite altimeters survey the entire global ocean, or very close to all of it, it is plausible that altimeter measurements can separate ocean volume changes from low-frequency redistributions of ocean mass with significantly shorter time series. Thus, even though less than a decade of precise altimetric data from T/P exists, these data might be as useful as much longer tide gauge series for certain problems.

We can extend this example further by imagining an ocean whose volume was indeed changing with time, but in which there were no dynamic redistributions of mass. Although this example is as artificial as the previous one, it will serve to illustrate a slightly different point. In this case, in principle just one perfect tide gauge would be adequate for determining the volume change rate. But this assumes, of course, that the volume change rate is the same everywhere on the planet. Unfortunately, models of the ocean volume change rate associated with an enhanced greenhouse effect (e.g. Russell *et al.,* 1999) suggest that this may not be the case. If these model simulations are correct, then even in our simple example where there are no mass redistributions to deal with, the relatively poor spatial sampling of the real tide gauge network can again lead to misleading conclusions. Obviously, the nearly complete global sampling afforded by satellite altimetry is a unique advantage for studies of very low frequency sea level change. No conceivable tide gauge network can provide the capabilities of a T/P-class altimeter satellite.

The foregoing examples were unrealistic in that the measurements, from both the tide gauges and the altimeter, were assumed to be free of error. Of course, each is subject to various errors (for example, Douglas in Chapter 3 discusses tide gauge errors in some detail; see also Pugh, 1987). A careful assessment of the errors in both is required before it will be clear how these observations should be combined to arrive at reliable estimates of very-low-frequency sea level change. This chapter focuses on satellite altimetry and considers the errors in those measurements, especially errors that take the form of a long-term drift. Obviously such errors are potentially fatal to the use of altimetric heights to measure real long-term trends in sea level. But the examples above do serve to show that there is great promise in using altimetric heights to make determinations of very-low-frequency sea level change.

6.1.2   History of Satellite Altimetry

NASA's satellite altimetry program was originally formulated at the 1969 Williamstown Conference (Kaula, 1970). An altimeter was actually used to scan the Moon by Apollo 14 (Kaula *et al.*, 1974). The first altimeter used for experimental Earth measurements was operated on the Skylab manned space station (McGoogan *et al.*, 1974). The first altimeter on an unmanned satellite was the Geodynamics Explorer Ocean Satellite 3 (GEOS-3), launched in 1975 (Stanley, 1979). Geos-3 provided useful data on eddy variability of the Gulf Stream region of the Atlantic ocean and demonstrated the potential of satellite altimetry for oceanography and marine geodesy (e.g., Douglas *et al.*, 1983). In 1978, the Seasat altimeter satellite was launched. In addition to the altimeter, it carried a microwave radiometer for correcting the altimeter measurements for delays caused by tropospheric water vapor. While Seasat's lifetime was only 3 months, it provided the first data sets sufficiently accurate to be used for global oceanographic studies (Lame and Born, 1982; Cheney *et al.*, 1983).

Geosat, launched in 1985, had performance similar to that of Seasat except it did not carry a microwave radiometer. It was operational for nearly 4 years and achieved many successes (see *J. Geophys. Res.* 95(C3) and 95(C10), 1990). In 1991, the European Space Agency launched the initial non-U.S. radar altimeter onboard ERS-1. It provided better performance than the Geosat altimeter, but still did not meet the requirements for regional or global sea level change studies as defined for the T/P mission.

T/P, launched in 1992, ushered in an entirely new era in satellite altimetry. Both the altimeter and orbit errors were only a few centimeters, giving sea level measurements accurate to 3–4 cm. These improvements were achieved by (1) making altimeter measurements at two different radio frequencies (Ku and C bands), which enabled the frequency-dependent ionosphere errors to be removed; (2) making simultaneous measurements of nadir columnar water vapor using a microwave radiometer for correcting the delay due to the wet troposphere; and (3) improving the orbit determination accuracy by using improved tracking techniques and geopotential models. Remarkably, T/P has been collecting measurements for 7 years as of this writing, well past its design life. The results shown in this chapter are entirely based on T/P, but a summary of the early sea level change results from Seasat, Geosat, and ERS-1 will be given to place the T/P results in a historical context.

A number of attempts have been made to measure global mean sea level variations from earlier satellite altimeter missions. Results using Seasat's 3-day repeat orbit showed 7-cm variations for estimates of global sea level over a month (Born *et al.*, 1986). Tapley *et al.* (1992) used 2 years of Geosat altimeter data and found 17-day values of variations in mean sea level with an RMS error of 2 cm and a rate of 0 ± 5 mm/yr. The largest errors were attributed to the orbit determination, ionosphere, wet troposphere delay corrections, and unknown drift in the altimeter bias (not independently calibrated for Geosat). Wagner and Cheney (1992) used a collinear differencing scheme and 2.5 years of Geosat altimeter data to determine a rate of global sea level rise of -12 ± 3 mm/yr. When this data was compared to a 17-day Seasat data set, a value of $+10$ mm/yr was found (Wagner and Cheney, 1992). The RMS error of the Geosat variations was still a few centimeters, even after the application of several improved measurement corrections. The ionosphere path delay correction was identified as the single largest error source, but there were many other contributions including errors in the orbit, wet troposphere correction, ocean tide models, altimeter clock drift, and drift in the altimeter electronic calibration. Since the Geosat study of Wagner and Cheney (1992), several improvements have been made to the Geosat altimeter measurement corrections (ionosphere, tides) and the orbit determination. However, Nerem (1995b) and Guman (1997) still find that Geosat mean sea level measurements are not of sufficient quality to allow a determination of the rate of mean sea level change accurate to 1 mm/yr, although the latter study succeeded in developing a tie between Geosat and T/P that allowed a reasonable estimate of sea level change ($+1.0 \pm 2.1$ mm/yr) over the decade spanning the two missions. Some investigators have examined ERS-1 altimeter data for long-term sea level change (Anzenhofer and Gruber, 1998; Cazenave *et al.*, 1998; Guman, 1997), but in general these results are less accurate because the ERS-1 altimeter is single frequency, and the orbits are less precise than T/P. Due to instrument problems, ERS-2 will be of limited use for studies of mean sea level change (Moore *et al.*, 1999). The Geosat Follow-On (GFO) mission has suffered from instrument problems since its launch in 1998, although the mission engineers are still optimistic that this satellite will one day produce useful altimeter measurements. Clearly T/P altimeter data represent the highest accuracy currently attainable from satellite altimetry.

6.1.3 Organization of this Chapter

The remainder of this chapter is organized as follows. First we present a brief review of results obtained to date on determining low-frequency sea level change from satellite altimetric data. We then review the altimetric measurement system in some detail for readers unfamiliar with the topic, trying to make estimates of the types of errors associated with the various component of the system. Next we turn to methods of detecting and correcting low-frequency trends in the altimetric data—an essential element in the application of altimetry to the determination of long-term trends, just as maintaining a consistent reference level is essential in the operation of tide gauges (Douglas, Chapter 3). Finally, we summarize the low-frequency sea level change results obtained with the T/P data, and conclude with an assessment of the prospects for improvement in the future.

6.2 THE ALTIMETER MEASUREMENT

6.2.1 Measurement Description

The first point to make about satellite altimeter data is that they are not simply the product of an instrument, but come from a *measurement system.* Several different observations need to be made to produce high-quality sea surface heights, and maintaining the quality of the final height observations requires high performance by all components of the system. As we proceed through this discussion of the various components, we will also briefly address the errors associated with each component, paying particular attention to whether a given component of the system might be prone to drift errors.

For the purposes of this discussion the system is separated into four basic components, illustrated schematically in Fig. 6.1. First there is the basic measurement of the distance from the satellite to the sea surface, which is called the satellite range and is so labeled on the figure. Second, to interpret this range as a height relative to some stable reference system, we also need to know the orbital height of the satellite in that reference frame, the zero point of which is labeled as the reference ellipsoid on Fig. 6.1. Third, one must realize that the satellite range is actually inferred from the travel time of a microwave radar pulse sent out from the altimeter, reflected from the sea surface, and then received again at the altimeter. To do the conversion from time delay to distance, we need to have an accurate value for the effective index of refraction along the path from the altimeter to the sea surface. Thus, the system requires measurements necessary to estimate the index of refraction of the atmosphere (troposphere and ionosphere). Fourth, since the altimeter's radar pulse actually samples an area of the sea surface that is relatively large (order of square kilometers), there is some ambiguity associated with how this average corresponds to mean sea level. Biases due to surface waves and the tidal phase must be considered in order to make a proper interpretation of the satellite range. These effects will be discussed together under the term surface effects. After discussion of these various components, this section will conclude with a brief summary of the errors in the overall system.

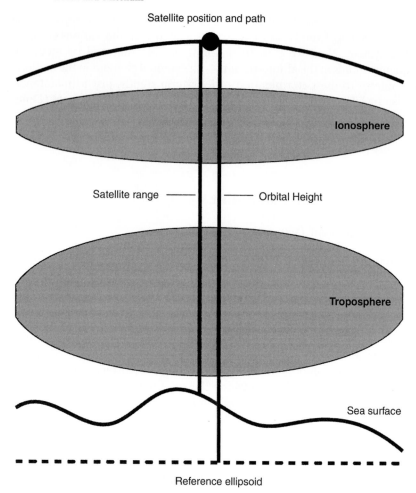

Satellite position and path

Ionosphere

Satellite range ———   ——— Orbital Height

Troposphere

Sea surface

Reference ellipsoid

Figure 6.1 Schematic of the measurement geometry for satellite altimetry. The satellite's instantaneous position is shown by a large solid circle, and the smooth curve passing through this point represents the satellite's orbital path. The instantaneous shape of the sea surface is shown by a solid curve at the bottom of the figure and the reference ellipsoid is shown as a dashed line. The shaded ellipses represent the ionosphere and the troposphere. Note that these portions of the atmosphere are not to scale.

Understanding how altimeter range measurements are collected and converted to sea level measurements is conceptually straightforward, although the details are complex. The satellite transmits a radar pulse toward the ocean surface. After passing through the atmosphere, the pulse arrives at the atmosphere/ocean boundary, interacts with the ocean, and is then reflected back toward the satellite, again passing through the atmosphere. If we ignore atmospheric effects on the pulse propagation and the interactions between pulse and the ocean surface, the range h can be found by measuring the difference between the time of pulse transmission and the pulse arrival back at the satellite, multiplying by the speed of light, and dividing by 2:

$$h = \frac{(t_A - t_T)c}{2}.$$ (1)

What the altimeter actually measures is the average waveform of thousands of returned pulses as a function of time. These short temporal averages are further averaged over 1-intervals for T/P, resulting in the basic 1-Hz data used by most T/P investigators. Since the satellite is moving at nearly 7 km/s, the T/P range measurement is actually an along-track average for approximately 6–7 km. Other satellite altimeters have slightly different averaging periods and along-track distances, but these values do not change much from one altimeter to another. Later during ground processing, the waveform is processed to yield the time of pulse arrival (usually taken as the midpoint of the leading edge of the waveform), the significant wave height (from the slope of the leading edge of the waveform), the radar backscatter coefficient (the amplitude of the returned waveform), and many other measurement variables. If the altimeter is dual-frequency (such as on TOPEX/POSEIDON, and the future Jason-1 and ENVISAT), then this procedure is done for measurements collected on both frequencies. For a more complete discussion of these basic measurement issues, the reader is referred to Stewart (1985).

Putting aside for now altimeter path length corrections due to the atmosphere, the precision of the measurement of the time delay is primarily limited by the number of altimeter pulses averaged in the basic time interval, 1 s for T/P, or equivalently in the basic along-track averaging length, 6–7 km for T/P. Averaging for longer periods results in more precise estimates, but results in less spatial resolution and introduces errors due to short length–scale sea surface height variability that is incorrectly averaged together into a single height estimate. Of course, accuracy is a somewhat different issue than precision. To obtain the time delay, the altimeter relies on an internal time reference, or a clock. If the clock is drifting, then even a precise measurement will be biased high or low. Such clock errors are a primary concern in the range measurement, because these errors can lead to low-frequency drifts in the range measurements.

In the design of the T/P altimeter (and most of its predecessors), it was realized that drifts in the timing system could be a serious source of error and an internal calibration system was included in the system (Hayne *et al.*, 1994). Estimates of the range stability—actually the range drift—are routinely produced by the T/P project and corrections applied to the final altimetric data provided by the project. Typically, these corrections are very low frequency in nature and have a peak-to-peak range during the T/P mission of order 10 mm. This internal calibration, however, has been brought in to question (Chambers *et al.*, 1998), and whether it is functioning as designed should still be considered an open issue. However, we know of no compelling reason to suspect the basic range measurement, (i.e., the basic time delay measurement) to be subject to systematic drift errors.

While we will later describe a number of atmospheric and environmental corrections to the range measurement given in Eq. (1), these are not necessary to understand how satellite altimeters can measure sea level. Note that the altimeter range measurement is just the altitude of the satellite above the ocean surface. It is not a measurement of sea level, but sea level can be computed with accompanying knowledge of the orbital position of the satellite.

6.2.2   The Satellite Orbit

The next part of the overall system to be discussed is the determination of the satellite's orbital height. Referring to Fig. 6.1 again, we see that without this independent measurement of the orbital height relative to some fixed coordinate frame, that is, relative to the reference ellipsoid, the range measurement would not be very useful. It would actually tell us more about the height of the satellite than the height of the sea surface! To determine the sea surface height relative to the reference ellipsoid, it is necessary to independently determine the height of the satellite, and difference the range h and orbital height R in order to obtain a sea level estimate. If the height of the satellite, R, can be determined in some suitable Earth-centered/Earth-fixed reference frame (relative to a reference ellipsoid that best fits the actual shape of the Earth), then the height of sea level, S, may be computed as

$$S = R - h. \tag{2}$$

There is an additional complication in that oceanographers would prefer that the reference surface be the oceanic geoid rather than an idealized ellipsoid, but for our purposes this distinction is not important, since we are mainly interested in temporal sea level changes and any reference surface that does not change with time is satisfactory. So how do we independently determine the height of the satellite relative to the height of the reference surface? Satellite geodesy techniques for orbit determination provide the solution.

Orbit determination can be described as a process that combines knowledge of the dynamics of Earth-orbiting spacecraft with very precise observations of the spacecraft. For T/P, measurements of the spacecraft range and/or range-rate are made from both Earth-based tracking stations (Satellite Laser Ranging (SLR) and the DORIS Doppler system) and space-based satellites (the Global Positioning System (GPS) constellation). Differences between the actual tracking observations and predicted observations based on the dynamics are used to adjust various dynamical (e.g., atmospheric drag, gravity) and measurement (e.g., tracking measurement biases) parameters to obtain better agreement. As of this writing, orbit determination methods are shifting from techniques that rely more on accurate modeling of the satellite dynamics (because the tracking data are geographically sparse) to techniques that rely more on the tracking measurements (because new tracking systems such as GPS provide virtually continuous orbital information).

The quality of the T/P precision orbit determination has been a major factor in the overall success of the mission. This success has come about because of several factors. First, the mission was designed with precision orbit determination in mind; thus the satellite is flown at a relatively high 1336-km altitude that reduces non-conservative forces, such as atmospheric drag, which are difficult to model. Second, significant improvements were made in our knowledge of Earth's gravity field, which is important because gravity is the dominant force acting on the satellite. Third, the satellite is tracked by several independent systems: SLR, DORIS, and GPS. Obviously this is a very complex activity and only a brief overview can be offered here. The reader interested in further information on precision orbit determination techniques is referred to Tapley *et al.* (1994b) and Nouel *et al.* (1994).

Basically, as the satellite moves about Earth, occasional observations of components of its position (e.g., range by SLR) or velocity (e.g., range-rate by DORIS) are made from ground stations, the positions of which are known accurately relative to the reference ellipsoid (Fig. 6.1). In addition to these point estimates of the satellite's height or speed, a physical model of the orbit is available that enables computation of a continuous estimate of the satellite's position and velocity. This physical model includes the effects of conservative forces (i.e, Earth's gravity field) (Nerem *et al.*, 1994; Tapley *et al.*, 1996) and non-conservative (e.g., atmospheric drag, solar radiation pressure) (Marshall and Luthcke, 1994) forces. The accuracy of the computed orbit is the most severely limited by our knowledge of Earth's gravity field, and to a lesser extent by the parameterizations necessary to model the non-conservative forces acting on the satellite. The model has adjustable parameters, however, that can be used to force the model orbit to conform to the observations provided at various points by the tracking stations. In a sense, we can view the precision orbit determination problem as obtaining a series of fixes on the satellite, and then interpolating between the relatively sparse fixes through the integration of the physical model for the satellite's orbit.

T/P also carries an experimental GPS tracking system (Melbourne *et al.*, 1994) which has produced orbits estimated to be accurate to 20–30 mm radially (Bertiger *et al.*, 1994; Schutz *et al.*, 1994; Yunck *et al.*, 1994). Because the GPS satellite constellation tracks T/P continuously in a three-dimensional fashion, the orbits have been determined using a "reduced-dynamic" orbit determination technique (Yunck *et al.*, 1994) which is less dependent on the dynamical models and thus less susceptible to errors in those models. However, only time periods when the anti-spoofing (A/S) security measure (Melbourne *et al.*, 1994) was not turned on have been processed, because the T/P GPS receiver is not equipped to compensate for A/S degradation. Since A/S-off has only occurred for roughly a dozen 10-day cycles, GPS cannot be used as a primary tracking technique for T/P. However, as the follow-on mission to T/P, Jason-1 will carry a more sophisticated GPS receiver that will work even when A/S is activated, and thus will be one of the primary tracking techniques for this mission.

A final orbit-related issue that warrants discussion is the effect of center-of-mass variations, since the center of mass of the earth (about which the satellite orbits) is the origin for satellite-based sea level measurements. The Cartesian coordinates of the center of mass of the earth with respect to a "crust-fixed" reference frame (defined by the location of the tracking stations) vary due to mass redistribution in the Earth system, primarily in the oceans and atmosphere. All satellites orbit the center of mass of the aggregate solid earth/ocean/atmosphere system. If the center of mass is the reference frame origin for the altimeter measurements of sea level, center-of-mass variations must be properly accounted for in the realization of the tracking station locations within this frame. These "geocenter" variations, which have been measured using precise geodetic satellites such as LAGEOS (e.g., Chen et al., 1999), are generally less than 20 mm. However, most of this variation is at seasonal periods, and significant secular trends in the geocenter components are expected to be quite small. Only recently have time series of geocenter variations been determined with accuracies approaching that needed for mean sea level studies, and thus these effects are not incorporated in current precision orbit solutions. While the effect is not expected to be large (due to averaging over the oceans and the accommodation of the error by other orbit parameters), the monitoring of the geocenter will likely be important for providing a well-defined origin for long-term measurements of sea level change (Nerem et al., 1998) when data from multiple missions are combined over several decades.

What sort of errors might we expect from the orbit determination? The spectrum of the orbit errors tends to have peaks once and twice per orbital revolution in addition to other lower frequency terms. Thus the orbit error tends to be large-scale in comparison with oceanic scales of variability. But over several days, during which time the satellite orbits Earth many times, the temporally varying components of these errors decorrelate (Marshall et al., 1995) and are not so serious for studies concerned with low-frequency variability. Very good estimates of the orbit error for T/P have been determined by comparing orbits computed from SLR/DORIS systems with GPS-determined orbits, which have largely independent sources of error. The RMS height error of the satellite as determined from these comparisons is ~20–30 mm (Marshall et al., 1995), which is fortunate because this is precisely the variable that needs to be well determined so that estimates of sea level trends not be biased by the orbit determination. A significant portion (~10 mm) of the RMS radial orbit error can be attributed to errors in the gravity field model (Nerem et al., 1994; Tapley et al., 1996); these errors are time-invariant at a fixed geographic location (if the ascending and descending tracks are processed separately) and thus do not affect the mean sea level measurements presented here. The remaining error (~10–20 mm) is primarily due to errors in modeling conservative forces, such as gravity effects arising from ocean tides. Measurement errors and reference frame definition (Nerem *et al.*, 1998) are smaller but nevertheless significant components.

6.2.3   Effects of the Atmosphere

In the preceding discussion of the satellite range measurement, it was mentioned that an estimate of the index of refraction along the path of the radar pulse is required. Two atmospheric effects that determine the actual speed of the radar pulse need to be considered in order for us to convert the time delay into an accurate range measurement. The two dominant effects are the ionosphere and troposphere shown schematically in Fig. 6.1. Those familiar with the structure of the atmosphere will note immediately that this drawing is not to scale, but is rather intended to indicate the importance of these portions of the atmosphere in the determination of the index of refraction. The ionosphere retards the pulse due to electromagnetic interactions with free electrons. In the troposphere we separate the effect into two parts, the dry correction that measures the dry-air mass the pulse must traverse, and the wet correction that takes into account the additional retardation due to water vapor in the troposphere.

Discussing the effect of the free electrons in the ionosphere first, we start by noting that the amount of retardation, converted to a change in the range measurement, is up to about 100 mm, with a typical value being of order 30 mm. This value, however, changes significantly with several variables. First, the free electrons are primarily due to the interaction of the solar wind with the upper atmosphere. Therefore there is a large difference between the day and night sides of the earth and thus over a single satellite orbit. Also, there is a significant change seasonally as the solar flux on a unit area changes. That is, there are fewer free electrons at higher latitudes than at low latitudes where the sun is more directly overhead. Finally, the electron content, and thus the correction to the range measurement, varies with the solar wind changes during the solar sunspot cycle—an effect that is well known to amateur radio operators.

Correcting for the time delay associated with the ionosphere requires knowledge of the total electron content in the path connecting the satellite to the sea surface. For so-called single frequency altimeters these estimates are provided by models of the ionosphere. We will not discuss these models, however, since the T/P mission, which carries a dual-frequency altimeter, adopts a different approach (Imel, 1994). With a dual-frequency altimeter, the radar observations of the time delay are done at two different frequencies. The reason for this is that the difference between these two measurements enables determining the ionospheric correction without an external model because the interaction with the free electrons has a known dependence on the frequency of the electromagnetic wave.

As the radar pulse leaves the ionosphere and traverses the troposphere we need to consider the wet and dry corrections to the range that were mentioned above. The dry correction is simply proportional to the total mass in the satellite to sea surface path and typically gives rather large range correction of order 2.3 m. The correction itself is computed from the surface atmospheric pressure taken from a numerical weather prediction model; this is appropriate since the surface atmospheric pressure at any point is very nearly a measure of the total mass over that point on the sea surface. Although the dry correction is rather large, its variation in time or space is typically less than 20–30 mm, because of the relatively small range over which the surface atmospheric pressure changes.

The wet tropospheric correction is quite different. This correction is due to the retarding effect of water vapor in the atmosphere, which is highly variable in space and time. There is more water vapor at low latitudes where the water is warmer; the range correction there is of order 300 to 400 mm. A more typical global value is 100 to 150 mm, since at mid-to-high latitudes the range correction is much smaller. But in both cases the day-to-day and season-to-season changes are comparable to these typical values and can therefore introduce significant errors if not properly accounted for. Unlike the dry correction case, numerical weather models are not adequate for doing this correction because the water vapor is both more variable than the surface pressure and less accurately computed by the models. To correct the range it is necessary to obtain independent measurements of the integrated water vapor. This is accomplished by flying a microwave radiometer along with the altimeter, as is done for T/P (Ruf *et al.*, 1994), ERS 1 and 2, and Seasat.

As before, we want to consider the types of errors that might be expected from these atmospheric corrections. In particular we ask whether low-frequency drift is likely. To our knowledge there is no evidence that the ionospheric correction from the dual-frequency altimeter on T/P is subject to low-frequency drift, although we cannot absolutely rule out the possibility. For the dry tropospheric correction, low-frequency drift could arise if the surface pressure fields from the numerical weather models used were drifting with time. This is not as likely as it might seem, however, since the weather models are constantly assimilating actual weather observations, including a great deal of *in situ* pressure data. This assimilation should keep the atmospheric pressure field in the model from drifting significantly. Also, since the variability associated with the dry correction is rather small, a drift should be relatively easy to detect.

The wet correction, depending as it does on another complex observational system, requires more careful scrutiny. In fact, a drift in the radiometer observations, and thus in the wet range correction, has already been detected and corrected during the course of the T/P mission. The possibility of such a drift was suggested by an analysis of the drift of the T/P height measurements relative to tide gauge measurements (Mitchum, 1998), and subsequently identified by Keihm *et al.* (1998). This will be discussed again in the next section of this chapter; for now we only note that the wet correction is a likely place to look for possible drifts in the overall T/P system.

6.2.4 Other Measurement Corrections

The final set of corrections to be discussed is a set of adjustments that must be made to properly interpret the satellite height measurement, which is computed from an average of many time delay estimates from within the satellite "footprint." This footprint is of order several tens of kilometers squared, and it is not difficult to imagine that the satellite average obtained once every 10 days or so can differ from the average that we would obtain if we had continuous *in situ* measurements of sea level over the entire area of the footprint at every instant in time rather than a single sample each time the satellite flies over. If there is such a difference we term this a surface effect, and there are three such effects that we will discuss. These effects are referred to as the sea-state bias, the tides, and the inverted barometer adjustment.

The first of these surface corrections, the sea-state bias, arises because the surface of the ocean is rarely perfectly smooth, but is instead nearly always covered by capillary waves and surface gravity waves. The satellite footprint is large enough that in principle these waves will average out to nearly zero over that spatial scale, but this does not work in practice. One reason for this is that the troughs of the waves reflect the radar energy more effectively than the wave crests, leading to a bias in the return to the satellite. That is, we get more reflection from wave troughs than crests and when the average is made it is found to be lower than the mean height that would be obtained if no waves were present. There are actually several distinct effects involving biases due to surface waves, but for simplicity we will combine these all under the term sea-state bias.

To do the best correction for this bias we would need to know the wave heights in the footprint as a function of the wavelength of the waves; that is, we need to know the complete wave spectrum. In fact, we can measure the wave height (SWH) from the T/P radar pulses, but the spectrum is not available to us. To obtain a proxy estimate of the effect of the wave spectrum, wind speed is also needed (Fu and Glazman, 1991) in an analysis of the sea-state bias. The idea is that when the wind is blowing, the waves tend to be short and choppy, while during calm conditions the dominant waves are relatively long swells from further away. The correction for the sea-state bias is thus parameterized in terms of the SWH and wind speed and the parameters of the model are obtained by fitting to the T/P data itself. The details of this procedure are beyond our scope here, but the interested reader is referred to Gaspar *et al.* (1994).

The ocean tides give rise to another surface effect that must be accounted for, but of a very different sort. The problem here is that the altimeter observes a given point on its ground track very infrequently, once every 9.9 days for T/P, in comparison to the dominant tidal frequencies, which are once or twice per day. Therefore, each time the altimeter passes over this point, the tidal signal is at a completely different phase, meaning that it may be high tide one time, low tide at another time, or anywhere in between. So we have an aliasing problem in which a signal of one frequency is not sampled often enough and ends up masquerading as a lower frequency signal. For example, the M_2 tidal component at a period of 12.42 hours appears in the T/P data at a period of about 62 days. For readers unfamiliar with the aliasing problem, a good discussion can be found in Bloomfield (1976). Aliasing is a significant problem for altimetry because the ocean tides have surface deviations of order 1 m, compared to a few centimeters for the low-frequency ocean signals that we wish to observe. If these aliased tidal signals could not be removed, it would be difficult, if not impossible, to make proper interpretations of the sea surface height time series obtained from the altimeter.

To deal with the tides it is necessary to predict the tides using a model and to remove this estimate from each of the altimetric height measurements. In practice, since the tidal alias periods are easily computed from the motions of the sun and moon and the rotation of the earth, the T/P data itself can be used to make empirical tidal models once long enough altimetric time series have been obtained. In addition, hydrodynamical tidal models can be used, especially if *in situ* and T/P data can be assimilated. These tidal modeling efforts have been extremely successful during the T/P mission, and the tide models derived from T/P are now the best available in large parts of the world's oceans. For more complete information about estimating and correcting for tides in the T/P data, see Molines *et al.* (1994).

The final surface correction to be discussed is called the inverted barometer correction. This should probably be referred to as an adjustment rather than a correction, however, since we are not correcting any error in this case. To understand what this adjustment is, consider that the altimetric observation is simply a geometric measurement of the height of the sea surface from the reference ellipsoid. What oceanographers often need, however, is an estimate of the pressure deviation due to the sea surface height's deviation from its long-term mean value. Such pressure estimates can be used to infer surface circulation, for example, from the geostrophic relationship. In estimating the pressure due to the water deviation, however, we need to adjust for any pressure fluctuation due to increased or decreased atmospheric load as measured by surface atmospheric pressure. If it is desired to make this conversion from sea surface height to ocean surface pressure deviations, one must add in the atmospheric pressure deviation, which is referred to as making the inverted barometer correction, or adjustment. In essence we obtain the pressure field that the subsurface ocean would experience if the atmospheric pressure field were constant. The atmospheric pressure fields used for this adjustment are the same as used for the dry tropospheric correction to the range discussed earlier.

Making an inverted barometer adjustment is not so relevant to studies aimed at determining sea level change, since sea level is also a geometric measurement and is therefore analogous to the altimetric sea surface height measurement without the inverted barometer adjustment applied. We do not make this adjustment in most of the work that we will present here, but it is used almost routinely in studies of ocean circulation. We do, however, apply this correction in some of our calculations of regional sea level change in order to reduce "noise" due to regional scale atmospheric variations.

In the case of the surface corrections, particular attention must be given to these errors since all of the corrections rely on model output rather than direct observations. The inverted barometer adjustment is particularly noisy because of numerical weather model errors in the surface atmospheric pressure fields, which is another reason we do not generally apply this correction in our work. The tide models are not perfect, but there seems little chance of these errors introducing low-frequency drift errors. The tidal signals and the tide model estimates both occur at discrete frequencies, implying that the tide errors do also. Since T/P was designed to keep large tidal components aliased to relatively short periods (less than 1 year), these errors average out when estimating very-low-frequency trends. So tide model errors are not considered a significant source of drift error for T/P, although this may not be true for other altimetric satellites where the orbital parameters, and the associated tidal alias periods, are not chosen as carefully. The sea-state bias models, on the other hand, are difficult to assess in terms of the potential for low-frequency drift and such a possibility has to be kept in mind.

6.2.5   Error Sources in Satellite Altimetry

In each of the preceding sections we have described the basic components of the altimetric measurement system, and have given a very brief assessment of the possibility that each of the components of the system might give rise to low-frequency errors, or drifts. In this section we will assemble more information about the errors in the overall system, again giving special attention to the likelihood that any of these errors could bias estimates of sea level trends computed from the altimetric heights. As discussed at the beginning of the chapter, these are the types of errors that would be most damaging to any attempt to estimate sea level change rates from altimetry. For each of the components of the system we will cite the initial geophysical evaluations of the T/P data, which have been summarized by Fu *et al.* (1994). For the orbit determination and the tides we will note more recent evaluations because significant improvements have occurred in these areas.

Starting with the range measurement, evaluations done with the first 2 years of T/P data indicated that the range errors were of order 2 cm, not including uncertainties in the correction for the atmospheric index of refraction. The long-term stability of the range measurement, as fixed by the internal calibration, is still an area of debate. The overall size of the internal calibration correction, however, indicates that the stability is good to at least several millimeters per year, a remarkable achievement even if it may not be quite up to the standard needed for estimating global sea level change.

For the orbit determination, the prelaunch requirement was that the orbit should be determined to about 10 cm. In fact, the T/P science teams charged with computing precise orbits had already exceeded this standard by the time of the launch; the initial estimates of the orbit errors were approximately 3.5 cm, with less than 2 cm of this in the time-independent error mentioned above. The computed orbits have continued to improve since then and present error estimates are on the order of 2–3 cm for a typical global value (Marshall *et al.*, 1995). This ongoing improvement of the quality of the orbital calculations is one of the true achievements of the T/P mission, and many of the scientific studies being undertaken with these data would not be possible had this not happened. As discussed above, it is not expected that the orbit errors will lead to significant drift errors, and there is no evidence of such errors to date. As mentioned earlier, reference frame stability could become a matter of concern over several decades and multiple missions, especially since the satellite tracking techniques (on which the reference frame is based) are continually evolving.

The situation with the atmospheric corrections, which essentially result in estimates of the effective index of refraction for the atmosphere, is a bit more complicated, particularly with regard to the wet tropospheric correction. The initial T/P evaluations led to estimates of 0.5 cm for the error in the ionospheric correction, 0.7 cm for the dry tropospheric correction, and 1.1 cm for the wet tropospheric contribution to the range correction. The drift error in the wet correction that we discussed earlier is, however, worrisome. Although to our knowledge the drift error is now corrected (Keihm *et al.*, 1998), it is a correction that obviously has the potential to produce significant trend errors and must therefore be watched closely in the future and in other altimetric missions.

Finally, for the surface corrections, the initial error estimates from the summary by Fu *et al.* (1994) were about 2 cm for the sea-state bias and 5 cm for typical global tide model error. As with the orbits, the quality of existing tide models was much better than anticipated before launch, and the models continue to improve. It is now estimated that the tide model errors are approximately 1–2 cm (Shum *et al.*, 1997) in the open ocean, and as we have already stated, it is unlikely that these errors will introduce significant bias into long-term sea level trend estimates. The same appears to be true for the sea-state bias estimation, although this cannot really be established yet, particularly if the estimates of wind speed and significant wave height were to exhibit unrealistic low-frequency trends.

One way to make an overall assessment of the precision and accuracy of the T/P system for producing sea surface heights is to compare these heights to sea level measurements from tide gauges. Although we cannot easily attribute any errors so observed to a particular component of the altimetric system, such comparisons do provide an important end-to-end assessment of the total T/P system. A comparison of this type was carried out by one of us (Mitchum *et al.*, 1994) shortly after the launch of the T/P satellite. Those results indicated that the overall precision of the T/P data was better than 4 cm at any given location using the highest frequency data available from the T/P system. Cheney *et al.* (1994) averaged the data to monthly time scales and obtained error estimates of order 2 cm, indicating that much of the error in the T/P system is uncorrelated temporally and therefore averages down like random noise. Even more encouraging is that the fact that in the period following these studies, the T/P data have improved, and similar calculations now show errors of less than 3 cm for the highest frequency data, which is almost a factor of 2 reduction in the error variance. An example of a T/P and tide gauge comparison is shown in Fig. 6.2, which shows the tide gauge record at Pohnpei in the central tropical Pacific and the analogous T/P-derived sea surface height series in the vicinity of the island. This comparison is better than the globally averaged result, but is not particularly unusual for island stations in the tropical parts of the oceans. As we will discuss in the next section, the quality of comparisons such as these allows us to make sensitive analyses of the errors in the T/P data and identify drifts smaller than 1 mm/yr.

6.3   CALIBRATING THE MEASUREMENT

6.3.1   Calibration Overview

Since we are particularly concerned with identifying low-frequency errors, or drifts, in the T/P data when using these data for estimating sea level change, it would be most useful if we had a method for specifically estimating drifts from independent data. In fact, techniques to do this have been developed using the tide gauge measurements referred to above, as well as by using other satellite altimeters and other types of *in situ* data. Although our focus in this chapter will be on the use of the global tide gauge network, we will briefly review these various other methods as well. Of course, before any of these methods can be trusted one must have some confidence that the data used as a benchmark, for example the tide gauges, are themselves stable over long time scales. We will address this issue as well for the tide gauges, and the reader is also referred to the discussion of tide gauge data in Chapter 3.

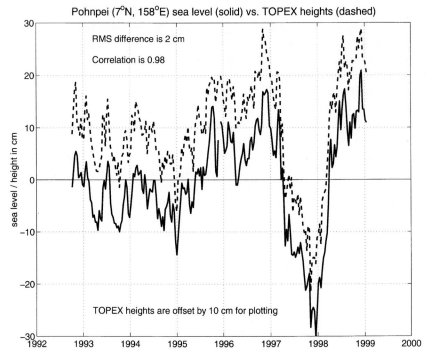

Figure 6.2 Comparison of the T/P heights (dashed line) and the tide gauge sea levels (solid line) at Pohnpei. The T/P heights are offset vertically by 10 cm to aid visualization. The T/P heights and the sea levels are from the improved signal matching algorithm described in Section 3.

Mitchum (1998) discusses various approaches to the problem of estimating the T/P drift error and categorizes the approaches that use *in situ* data according to the strategy adopted for minimizing the error of the drift estimate. This discussion can be summarized as follows. First, all of the methods proceed by computing a difference of the T/P heights with an *in situ* estimate of sea level. These *in situ* estimates can come from a single calibration site such as the offshore California Harvest oil production platform (e.g., Christensen *et al.,* 1994), from tide gauges in lakes (Morris and Gill, 1994), from dynamic height computed from open ocean temperature profiles, or from the global tide gauge network (Mitchum, 1994; Mitchum, 1998). Computing differences is done in order that the ocean signals that are common to both the T/P and *in situ* data will cancel, isolating the errors for further analysis.

Regardless of the source and extent of the *in situ* data, in each case there are one or many locations where differences are computed but only some fraction N of these can be considered statistically independent. For simplicity in this illustration, we assume that at each location the standard deviation of the (T/P minus *in situ* differences) can be characterized by a single value that we will denote as σ. Under these assumptions the variance of the global mean difference at any point in time, which results in a time series for the drift estimate, will be σ^2/N. The various approaches to the drift estimation problem can be classified according to the strategy adopted for minimizing this variance.

In the case of the single site calibration, such as is done at the Harvest platform, the strategy is to obtain the smallest difference between the T/P data and the *in situ* data, which are from a tide gauge, and thus to minimize σ^2. In the case of the global tide gauge analysis, larger σ^2 values are accepted, but N is made as large as possible. A similar strategy characterizes the approach using dynamic height from temperature profiles, although the σ^2 values will be larger due to the more indirect method of estimating sea level. The use of tide gauges only in the Great Lakes is an intermediate approach, wherein σ^2 is larger than at Harvest but smaller than at many open ocean tide gauges, but N cannot be made as large as with the open ocean approach. To date, the global tide gauge approach has arguably yielded the most sensitive results for the time-dependent error because the increase in N has outweighed the slightly larger values of σ^2 at the various gauges used in the analysis. The global tide gauge approach successfully identified an algorithm error in the early T/P data and also gave the first indication of a possible drift in the wet correction (Mitchum, 1998). The Harvest platform analysis, on the other hand, being the only technique that can presently detect mean, or time-independent, biases is extremely valuable for determining offsets between different satellite missions. It is important to note, however, that efforts to improve all of these methods are active at present, and results should improve with time. In summarizing the global tide gauge approach to estimating T/P drift we will give a short review of the development of the method, and also present the most recent results from this approach. But first we will briefly consider the stability of the tide gauges themselves, since assuming that these instruments are stable is a key assumption of the method.

6.3.2   Using Tide Gauges for Calibration

One of the advantages of using tide gauge data for calibration is that methods have long been in place to maintain the stability of the vertical reference point for tide gauges. Some discussion of these methods is given in Chapter 3, and more details are available elsewhere (e.g. Pugh, 1987). One of the key components of the tide gauge system is the tide staff, which is simply a calibrated pole placed close to the tide gauge and read periodically by a human observer. Comparing gauge and staff readings allows the tide gauge operator to identify drifts in the mechanical tide gauge system, since the very simple and direct staff measurements, although noisy, should not be subject to serious drift problems. But these staff measurements have often been questioned, and an illustration of the quality of the staff readings will be given here to provide some confidence in these measurements.

Our example is drawn from the Yap tide gauge station in the western tropical Pacific. This station was chosen because we have over 15 years of contemporaneous gauge and staff readings where the staff readings have never been used to correct the tide gauge data, thus keeping the two data sets completely independent. The staff readings are taken several times a week and do not resolve the ocean tides, so the staff heights were fit to a simple tide model using a least-squares criterion. The residuals from this tide fit were then averaged to monthly values. These monthly values are compared to the monthly mean gauge readings in the upper panel of Fig. 6.3, which show that there is very good agreement between these two time series at low frequencies. This agreement can be quantified as a function of frequency by computing a cross spectral analysis between these two time series, the results of which are shown in the bottom two panels of the same figure. The response function measures the ratio of the amplitude of the signals observed by the staff to those observed by the gauge. At periods longer than about a year, this function is very near unity, meaning that the staff successfully recovers the amplitude

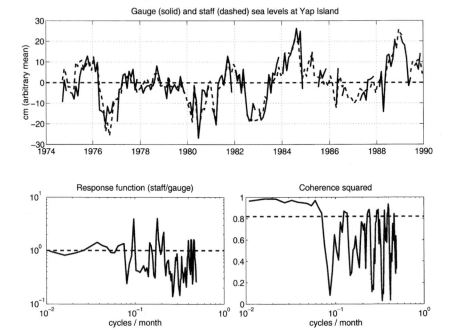

Figure 6.3 Comparison of sea level from the tide gauge (solid line) and the tide staff (dashed line) at Yap Island. The upper panel shows the monthly mean time series, and the lower two panels are from a cross spectral analysis of these two time series. At the lower left is the response function (staff variance to gauge variance); at the lower right is the coherence-squared function. The dashed line in the coherence-squared panel gives the 95% probability point for the null test that the true coherence-squared is different from zero.

of the sea level variability at those periods. In the lower right panel, the coherence squared function, which is basically a correlation squared as a function of frequency, is plotted, and at the same subannual frequencies the coherence values approach a value of 1, indicating nearly perfect agreement. The dashed line on this panel shows the critical coherence value such that values larger than this can be considered significantly different from 0 at the 95% confidence level. From the analysis it is clear that the staff measurements, despite being noisy and infrequent are faithfully recording the sea level variations, and with a system that is very unlikely to be drifting. Although this analysis at one station is not definitive, it gives some confidence that the use of tide staffs in maintaining the low-frequency stability of tide gauges is a viable and reliable approach.

There is, however, a much more serious problem with the tide gauge data that the staff data cannot address. Both the tide gauge and the staff measure sea level as a difference of the ocean height and the land on which the instruments sit. The stability of these instruments is monitored relative to the land near to the gauge by periodic surveying and in principle one can determine if there is instability in the land very close to the gauge. If the land is moving vertically on a larger scale, however, none of these measurements will detect it. This means that the sea level trend computed from tide gauge data must be considered to be the difference between the ocean sea surface trend and the vertical land motion rate. And thus a trend in the difference between the T/P data and the sea level from a given tide gauge must be considered as the sum of the T/P drift rate and the land motion rate.

A striking demonstration of the effect of land motion has been given by Cazenave *et al.* (1999) at the island of Socorro off the west coast of Mexico using independent estimates of the land motion at the site from DORIS measurements. For the purpose of demonstrating the problem here we have plotted the long time series of sea level obtained at two locations along the Hawaiian Ridge in the central subtropical Pacific. These locations are at Hilo on the south end of the main Hawaiian Island chain, and Honolulu at approximately the middle of the main Hawaiian Islands (Fig. 6.4). It is clear at a glance that the sea level rise rates at the two stations are significantly different even though the stations are quite close together. The interpretation of these results is that the Hilo tide gauge, which is located in an area of active volcanism, is subject to larger land motion than the gauge at Honolulu. Other examples of this sort are easily found, as discussed by Douglas in Chapter 3. The only ultimate solution to the land motion problem is to have space geodetic measurements, such as DORIS or GPS, at any tide gauge used in the drift estimation. But at present very few gauges are so equipped, and alternative methods for estimating land motion and assessing the uncertainty due to land motion have been devised. These methods will be briefly summarized at the end of this section, but first we will review the development of the basic methods used to estimate T/P drift from the global tide gauge network.

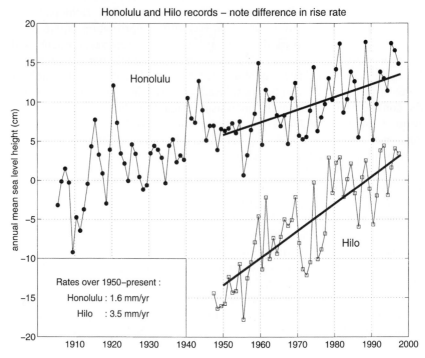

Figure 6.4 Sea level trends at Honolulu and Hilo in the Hawaiian Islands (after Nerem and Mitchum, 1999). The time series are annual mean sea levels and the rates quoted are from standard least-squares fits. The Hilo station is in an area of active volcanism.

The initial suggestion of using the tide gauge data as an assumed stable check for altimetric heights was given by Wyrtki and Mitchum (1990) in an analysis of Geosat altimeter data. This was followed by a more sophisticated, but still preliminary application of this approach to the first 2 years of T/P data (Mitchum, 1994). As the length of the T/P time series increased, the method was improved significantly. The discovery in the T/P data of an algorithm error that was successfully captured by the tide gauge analysis (Mitchum, 1998) resulted in a much wider acceptance of these estimates of the T/P altimeter stability. The method outlined by Mitchum (1998) matched T/P data to tide gauge data by simply choosing the T/P data at the nearest point of approach at the latitude of the tide gauge. Also, only the nearest four ground tracks were examined for each tide gauge. But a careful statistical analysis was done to assess which T/P and tide gauge difference series could be considered independent, and to determine an optimal way in which to combine the difference series from different tide gauges together in order to arrive at a globally averaged difference curve that can be interpreted as the T/P measurement drift rate. The problem with land motion was recognized, but at that time only a crude estimate of the magnitude of the potential bias error due to land motion was made. It was estimated that the net effect of land motion after averaging over the global network was of order 0.4 mm/yr, but with a larger uncertainty of order 1 mm/yr. This term in fact dominated the error estimate for the T/P drift rate. The method derived in the Mitchum (1998) paper results in the drift series shown in the top panel of Fig. 6.5. Subsequent improvements made to the method will be addressed now, and are the topic of a paper now being prepared for publication.

There were three obvious ways in which the method of Mitchum (1998) could be significantly improved. First, that paper raised the issue of a possible drift in the wet troposphere correction. Such a drift was subsequently identified by Keihm *et al.* (1998) and a modified correction proposed. This modified wet correction was incorporated into all of the subsequent estimations of the T/P drift. Second, the method that Mitchum (1998) used to match the T/P data to the tide gauge data was not necessarily optimal for eliminating ocean signals because there was no allowance for spatial or temporal lags. Third, no attempt to correct for land motions was made at all.

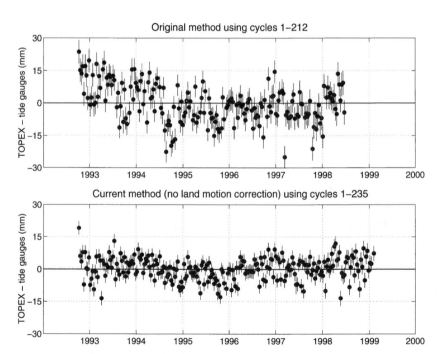

Figure 6.5   T/P drift estimates from the original method (upper panel) given by Mitchum (1998) and from the improved method outlined in Section 3. Each point is an average over the global tide gauge network during a 9.9-day T/P cycle, and the vertical bars are the error estimates for each cycle estimate of the drift series.

The second possible improvement—making a better ocean signal cancellation by matching up the T/P and tide gauge data more precisely—is necessary because the T/P and tide gauge data are not taken at precisely the same location. Since the T/P ground track spacing can be nearly 300 km, there can be significant temporal lags between the two time series in areas where the ocean signals are dominated by slowly propagating ocean waves. An example of this was given by Mitchum (1994) at the tide gauge station at Rarotonga. In such cases lagging the series appropriately results in a better cancellation of these ocean signals, and consequently a smaller variance in the differences, which in turn allows a more sensitive determination of the T/P drift rate. Similarly, simply taking the T/P data at the same latitude as the tide gauge, as was done by Mitchum (1998), does not allow for spatial deformations in the ocean signals. The solution to both of these problems is to define the "best" T/P data to use with the tide gauge sea levels for given temporal and along-track spatial lags, with these lags chosen so that the variance of the difference time series is minimum.

Making this modification to the original method and using the modified wet correction result in the time series shown in the bottom panel of Fig. 6.5. Comparing this series to that in the upper panel, which uses the original method, shows that the magnitude of the linear trend is somewhat reduced. This is due to the inclusion of the modified wet correction. In addition, the point-to-point variability in the time series is much smaller on the lower panel as compared to the upper. This reduction is from the improved cancellation of the ocean signals due to the better matching of the T/P and tide gauge time series. The reduced noise level in the drift series opens up the possibility that smaller trends might be detected and also that shorter term errors might be more reliably detected, providing a generally more useful tool for studying the T/P drift characteristics.

Figure 6.5 shows that improving the wet correction and allowing for spatial and temporal lags in the ocean signals observed by the two measurement systems significantly improve the drift estimate. There is, however, still the larger problem of what to do about land motion at the gauges and how to estimate the uncertainty in the T/P drift estimates due to these motions. As stated earlier, the ultimate solution to this problem awaits the availability of long time series of space geodetic information at each gauge, but in the interim alternatives have been devised and are continuing to be improved. In brief, at each gauge two estimates of land motion are made, one from the tide gauge data and one from the nearest geodetic measurements by either GPS or DORIS. Uncertainties for each of these estimates are also evaluated and the two are then combined to make the best estimate possible for each tide gauge. The estimate based on the geodetic information is simply the land motion rate at the nearest continuous GPS or DORIS station within 50 km of a coastline. Of course, for the majority of stations the closest station is many hundreds of kilometers distant. If no geodetic rate was available within 500 km, then no geodetic estimate was attempted. For the other stations, the land motion rate estimate was set to the observed rate, but the uncertainty was inflated with a function that increased with distance from the tide gauge. For stations more than approximately 200 to 300 km away, the uncertainties become large enough that the geodetic rates essentially have no weight in the analysis.

The internal estimate of the land motion rate—the one derived from the tide data—is somewhat more complicated. In this case it is assumed that the sea level change rate computed from long time series of sea level, which in this case means at least 15 years and usually more like 25 to 40 years, is approximately the true ocean rise rate minus the land motion rate. Since we are forced to make these estimates from relatively short time series (see Chapter 3 by Douglas for warnings about attempting this calculation), we also tried to estimate interannual ocean signals associated with the El Niño/Southern Oscillation events when computing the rate estimates. Under this assumption, a land motion rate estimate and its uncertainty can be made from the apparent rate and its uncertainty and from an estimate of the true background rate, which for our present purposes is taken to be 1.8 mm/yr (Douglas, 1991; Chapter 3). Clearly this calculation is somewhat speculative, but since our hope is to remove the bias from the drift estimate only when averaging many stations together rather than to accurately estimate rates at individual stations, we expect that the results will still be an improvement over no land motion correction at all.

Although this discussion of the land motion correction is brief, it does give the basic idea. When these land motion corrections are applied, we obtain the drift time series shown in Fig. 6.6 in the upper panel. To get an idea of the net effect of applying these land motion estimates at the individual gauges, we differenced this time series from the previous calculation that included the improved wet correction and the better signal matching but did not make any adjustment for land motion (Fig. 6.6, bottom). The net result, not surprisingly, is mainly a change of the estimated trend. The sign of this change corresponds to net subsistence of the land at about 0.5 mm/yr when all of the global gauges are averaged. Because we have had to make a number of assumptions in creating this land motion correction, it is necessary to ask how sensitive the results are to these assumptions. The critical assumption turns out to be the rate taken for the true ocean rise rate. If this parameter is varied, the inferred T/P drift rate changes significantly (Fig. 6.7), and for a true rate in the range of 1–2 mm/yr, the range of the T/P drift rate estimates is approximately 0.8 mm/yr. From this we conclude that the uncertainty due to the land motion is of order 0.4 mm/yr. This is still the largest uncertainty in the drift estimate, but is a significant improvement over the previous estimate (Mitchum, 1998) of order 1 mm/yr.

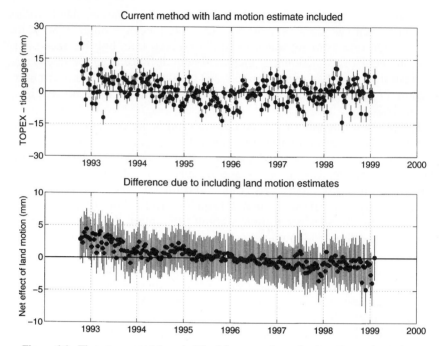

Figure 6.6 The upper panel is as in Fig. 6.5, except that a land motion estimate has now been made for each tide gauge time series. See the text for the details on how this land motion correction is estimated. The lower panel is the difference between the upper panel of this figure and the lower panel on Fig. 6.5, which shows the effect of adding the land motion estimates. The error bars in the lower panel are computed under the assumption that the errors in these two times series are not independent.

6.4   RESULTS FROM THE TOPEX/POSEIDON MISSION

6.4.1   Overview of the Achievements of TOPEX/POSEIDON

There is a rich collection of journal publications describing the scientific results from the TOPEX/POSEIDON mission. The initial results are described in two special issues of the *Journal of Geophysical Research:* "TOPEX/POSEIDON: Geophysical Evaluation," **99**(C12), 24,369–25,062, 1994, and "TOPEX/POSEIDON: Science Results," **100**(C12), 24,893–25,382, 1995. The former collection of papers contains overviews of the mission, gravity model development, orbit determination, and evaluations of data quality and accuracy, while the latter focuses more on scientific problems such as ocean tides, the mean sea surface, global sea level change, sea level variability, and ocean circulation. Another major scientific accomplishment not covered in these special issues is the monitoring of ENSO events using TOPEX/POSEIDON altimetry (Chel-

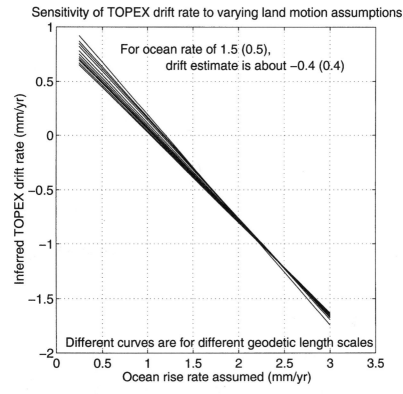

Figure 6.7   A sensitivity test of the method used to assign land motion estimates to the tide gauge sea levels. There are two parameters in the land motion corrections: the true ocean rise rate and the length scale over which the variance in the geodetic estimates of the land motion rate increases. The T/P drift rate was estimated using a wide range of these parameters and these rates are plotted versus the assumed ocean rise rate. The different lines on the figure correspond to different values for the length scale for the variance increase of the geodetic rate. The technique is sensitive to the ocean rise rate chosen, but relatively insensitive to the geodetic length scale chosen.

ton and Schlax, 1996). Here we provide an overview of only the results on estimating global mean sea level variations using TOPEX/POSEIDON altimetry. There have been numerous papers describing the global mean sea level variations observed by T/P (Cazenave *et al.*, 1998; Minster *et al.*, 1995, 1999; Nerem, 1995a,b; Nerem *et al.*, 1997a, 1999) and their spatial variation (Hendricks *et al.*, 1996; Nerem *et al.*, 1997b). Altimeter measurements have also been used to monitor mean water level in semienclosed or enclosed seas (Cazenave *et al.*, 1997; Larnicol *et al.*, 1995; and Le Traon and Gauzelin, 1997 among others) and lakes (e.g., Morris, 1994; Birkett, 1995) and rivers (e.g., Birkett, 1998), although these applications are outside the scope of this chapter.

6.4.2   Measuring Changes in Global Mean Sea Level

For the purposes of completeness, a summary of our processing of the T/P data for the determination of global mean sea level variations will be presented. While the T/P data have been processed in a variety of different ways (Cazenave *et al.*, 1998; Minster *et al.*, 1995, 1999; Nerem, 1995a,b; Nerem *et al.*, 1997a, 1999), the results are all quite similar. The data processing employed for the results presented here is essentially identical to that used in Nerem (1995a,b, 1997a) and thus will not be reproduced in detail.

In our processing, mean sea level variations are computed every 10 days by using equi-area-weighted averages (Nerem, 1995b) of the deviation of sea level from a 6-year along-track mean (1993–1998) computed exclusively from the T/P data (as opposed to using a more general multimission mean sea surface). Although T/P cannot measure "global" mean sea level because it covers only $\pm 66°$ latitude, tests have indicated the mean sea level estimates are very insensitive to this gap in latitudinal coverage (Minster *et al.*, 1995; Nerem, 1995b).

All of the usual altimeter corrections discussed earlier in this chapter (ionosphere, wet/dry troposphere, ocean tides, sea state, etc.) have been applied to the data, with one exception. No inverted barometer (IB) correction was applied to these data (Nerem, 1995a,b), because we were concerned about errors in the IB correction (Fu and Pihos, 1994, Raofi, 1998). Although improvements in the IB correction have been made (Dorandeu and Le Traon, 1999; Raofi, 1998), we argue it is the total sea level change signal that is of interest, and not its IB-corrected equivalent. If there was a secular change in mean atmospheric pressure, resulting in a secular change in mean sea level, we do not want to remove this signal from the results. Nevertheless, the IB contribution to secular changes in mean sea level over the T/P mission is less than 1 mm/yr (Dorandeu and Le Traon, 1999). Since the CSR 3.0 ocean tide model was used (which was developed with IB-corrected T/P data), not applying the IB correction introduced a slight error with a 58.7-day period associated with the S_2 atmospheric pressure tide normally modeled in the IB correction.

Modified geophysical data records (MGDRs) covering Cycles 10–233 from both the TOPEX and POSEIDON altimeters have been used in this study. Cycles 1–9 were omitted because they are suspect (Nerem, 1995b). The single-frequency POSEIDON altimeter does not provide an ionosphere correction directly, but since this altimeter is used in only about 10% of the mission, the somewhat lower accuracy of the DORIS-derived ionosphere correction (Minster *et al.*, 1995) does not significantly affect the results presented here. The MGDRs differ from the GDRs (used in earlier studies) in that an improved EM bias algorithm is employed (Gaspar *et al.*, 1994), and updated estimates of the σ_0 calibration were used (Callahan *et al.*, 1994). The on-board TOPEX altimeter internal calibration estimates have also been applied (Hayne *et al.,* 1994) and are designed to measure changes in the instrument calibration using measurements from a calibration loop in the instrument electronics. In addition, a correction of 1.2 mm/yr over 1993–1996 (flat from 1997 to present) has been applied for an apparent drift in the TOPEX microwave radiometer (TMR), which provides the wet troposphere correction (Keihm *et al.,* 1998). The latest improved orbits using the improved JGM-3 gravity model (Tapley *et al.,* 1996) have been employed. While some data editing is performed (shallow water, outliers, high mesoscale variability, etc.), Nerem (1995b) and Minster *et al.* (1995) have shown the mean sea level estimates to be very insensitive to this editing.

Figure 6.8 shows the cycle-by-cycle (10 days) estimates of global mean sea level for Cycles 10–233 computed using the techniques described in Nerem (1995b). The RMS variation of the mean sea level is roughly 7 mm after removal of a trend. Spectral analysis reveals much of the variability lies near periods of 511, 365, 182, and 59 days, all of which have amplitudes of 2 mm and greater. The variability at a period of 59 days is near the fundamental period at which the T/P orbit samples semidiurnal varying phenomena, such

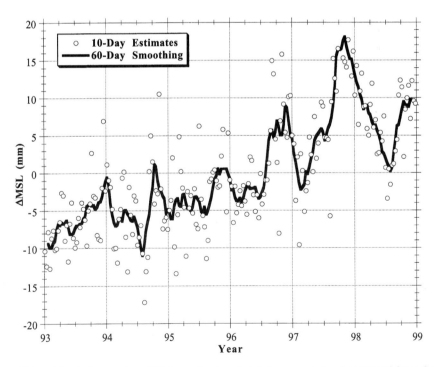

Figure 6.8   A time series of 10-day estimates of global mean sea level computed from the TOPEX/POSEIDON mission altimeter data and the same results after smoothing using a 60-day boxcar filter.

as atmospheric and ocean tides (such as the aforementioned S_2 error) and the ionosphere. This variability can be significantly reduced by smoothing the mean sea level values using a 60-day boxcar filter. As shown in Fig. 6.8, this reduces the RMS variation of the time series to less than 4 mm after removal of a trend. Also note the large 15- to 20-mm increase in mean sea level at the end of the time series, which we will show later is related to the 1997–1998 ENSO event (Nerem *et al.*, 1999). The power of satellite altimetry to spatially map the sea level change signal will also be reviewed later.

6.4.3   Calibrating the Mean Sea Level Measurements

We now want to correct the T/P mean sea level time series (Fig. 6.8) using the tide gauge instrument calibration time series (Fig. 6.5) to get our final estimate of the rate of sea level change. Note that we prefer to identify altimeter measurement error detected by the tide gauges, and directly correct the altimeter data, as done in the past for the oscillator correction (Nerem, 1997) and the wet troposphere correction (Keihm *et al.*, 1998). For this discussion, we also remove annual and semiannual variations from the time series because there is some evidence that the tide gauge calibration might have small errors at this frequency (the tide gauges are more sensitive to localized seasonal variations in heating, freshwater flux, and coastal trapped waves (Mitchum, 1998). Figure 6.9 shows the final time series of global mean sea level change from T/P. A simple least-squares linear fit to these data give a rate of +2.5 mm/yr with a formal error ±0.25 mm/yr (based on the scatter of the fit). We use formal error here because we cannot assemble a realistic error budget for global mean sea level otherwise, since we need to know the globally integrated errors due to mismodeling of the orbit, ionosphere, troposphere, tides, and so on. The formal error assumes the point-to-point measurement errors (one 10-day estimate of mean sea level to the next) are uncorrelated. If we account for the autocorrelation of the fit residuals (e.g., Maul and Martin, 1993), this increases the formal error to 0.4 mm/yr. However, we also corrected the T/P data using the tide gauge calibration values (to account for measurement errors that are otherwise inaccessible to us), and the estimated error in these values needs to be added to the formal error. Our estimate of the error in the tide gauge determination of the instrument drift is ±0.6 mm/yr, which is dominated by the land motion errors. The root-sum-square of the formal error and the instrument drift error gives a total error of 1.3 mm/yr. Our final rate estimate of +2.5 ±0.7 mm/yr (Nerem *et al.*, 1999) is valid only over the 6 years covering 1993–1998; we really cannot say if it is representative of the rate of long-term sea level change (as we will show later, the effects of ENSO variability significantly affect the rate estimate). Note that in any case, this estimate of sea level rise over 1993 to 1998 is statistically indistinguishable from the long-term rate determined from the tide gauge data (Douglas, 1991).

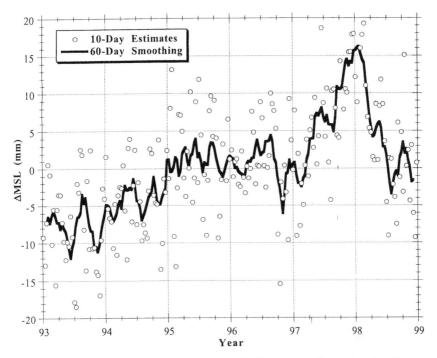

Figure 6.9   Same as Fig. 6.8, but after correction for instrument effects using the tide gauge calibration time series (Fig. 6.5) and removal of annual and semiannual variations.

6.4.4   Comparison to Sea Surface Temperature Changes

When studying measurements of sea level change, we often examine measurements of ocean temperature to determine if the observed changes in sea level are due to thermal expansion. However, measurements of temperature over the water column can only be made with *in situ* instrumentation (hydrographic measurements from ships, buoys, sondes, etc.), which have limited spatial extent. Sea surface temperature (SST) can be observed from space, using either infrared or microwave techniques, thus making SST a natural variable for comparison to the global measurements of sea level collected by T/P. Unfortunately, SST measurements reveal nothing about the temperature of the entire water column, and thus some assumptions (mixed layer thickness and temperature, thermal expansion coefficient, etc.) must be invoked to directly compare these measurements to sea level. For this discussion, we will qualitatively compare the observed spatial and temporal variations of sea level and SST, but need to remember that they need not be in agreement if the SST is not representative of the temperature of the mixed layer or if phenomena other than thermal expansion of the mixed layer are driving the sea level change.

We use a global $1° \times 1°$ SST data set covering 1982–present compiled at weekly intervals by Reynolds and Smith (1994), which is based on thermal infrared images collected by the advanced very high resolution radiometer (AVHRR) on board the NOAA satellites and is calibrated with *in situ* data. Using equiarea weighting, we can compute variations in global mean SST since 1982. This is dominated by a large annual change in global mean SST as described by Chen *et al.* (1998), which is initially surprising since the annual change in global mean sea level is only 2–3 mm. Chen *et al.* (1998) and Minster *et al.* (1999) reconciled this difference by observing that annual changes in continental water storage effectively cancel out the annual sea level change caused by changes in SST.

For the present discussion, we will simply remove the annual and semiannual SST variations from the original data set, and compute the global mean SST variations as shown in Fig. 6.10. In addition, we apply 60-day smoothing to be consistent with the smoothing of the sea level time series. A large increase in global mean SST is observed for each major ENSO event since 1982, with the largest change (0.35°C) occurring during the 1997–1998 event. Also shown in Fig. 6.10 is the variation in global mean sea level from T/P, after correcting for instrument effects using the tide gauge calibration (Fig.

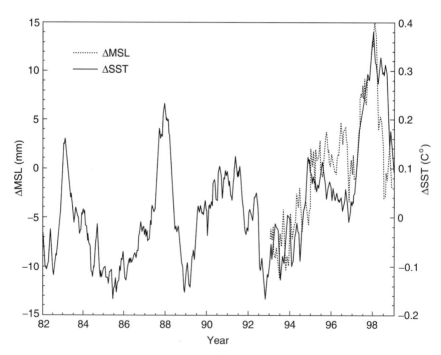

Figure 6.10 Comparison of global mean variations of sea level (Fig. 6.9) and sea surface temperature (Reynolds and Smith, 1994), after removal of annual and semiannual variations.

6.5), removing seasonal variations, and applying 60-day smoothing. Qualitatively, the comparison is quite compelling. Note that the increases in sea level and in SST at the beginning of 1997 occur nearly simultaneously, while the decrease of SST in 1998 lags that of sea level by several months, suggesting that temperatures in the subsurface returned to normal before the surface temperatures. These results suggest that the rise and fall in sea level during 1997–1998 were directly related to the ENSO event, and we will further demonstrate this by examining the spatial variation of these changes in the next section.

6.4.5   Spatial Variations

We can gain further insight into the cause of the observed mean sea level variations by examining the spatial variations in sea level change that contribute to the global mean. While variations in global mean sea level have great scientific and public interest as a barometer of environmental change, the true scientific and social implications will be determined by the spatial patterns of sea level change. As an example, anthropogenic-induced climate change, while likely to cause a rise in globally averaged sea level, will actually cause sea level to decline at some locations and rise in others (Russell *et al.*, 1995). This type of information will be critical for planning mitigation efforts related to sea level change. In addition, the spatial pattern of sea level change, if it could be mapped, provides another powerful constraint on climate change models by mapping the geographic "fingerprint" of the change. If the spatial variation of sea level change could be mapped using satellite altimetry and a similar pattern were identified in the output of the global climate models, this would be a powerful corroboration of the validity of the models.

Clearly, the true power of satellite altimetry lies in its ability to map the geographic variation of sea level change. As an example, T/P was the first altimeter mission to accurately map the seasonal variations in sea level, as shown in Fig. 6.11 (see color plate). There are many different spatio–temporal analysis techniques that may be employed to extract the spatial variability of sea level change from the T/P data records. A relatively simple technique for mapping the geographic variation of long-term sea level change is to compute the linear trend of sea level at each geographic location, as shown in Fig. 6.12 (see color plate) for the T/P mission. These trends were determined via a least-squares fit of secular, annual, and semiannual terms at each location along the T/P groundtrack. The problem with this technique is that sea level change at a single location often has considerable deviations from a linear trend, and these deviations tend to corrupt the trend estimate unless a long time series is available. Currently, the trends during the T/P mission are dominated by variability from recent ENSO events, which are clearly manifested in the satellite-observed sea surface temperature results (Reynolds and Smith, 1994) of the same time period (Fig. 6.12), as well as in numerical ocean models (Stammer *et al.*, 1996). However, as the sea level record from satellite altimetry lengthens, the ENSO variations will gradually average out, allowing the detection of climate signals. However, decadal variability may limit this approach until many decades of data are available.

More quantitative results can be obtained using statistically based analysis methods. The method of empirical orthogonal functions analysis (EOFs, also called principal component analysis) identifies linear transformations of the data set that concentrate as much of the variance as possible into a small number of variables (Preisendorfer, 1988). This method has been used in a variety of oceanographic and meteorological analyses to identify the principal modes of variability. If the spatial and temporal characteristics of a mode can be used to identify a physical cause, then the method can be a very powerful data analysis tool. Several investigators have used EOF techniques to help isolate the cause of the sea level change signals observed by T/P (Hendricks *et al.*, 1996; Nerem *et al.*, 1997b). Unfortunately, EOFs have so far been unable to separate ENSO-related variations from long-term sea level change, possibly because these processes may be interrelated (Cane *et al.*, 1996; Trenberth and Hoar, 1996). We will nevertheless review the EOF results here because they provide considerable insight into the cause of the mean sea level variations observed by T/P.

We start by converting the raw T/P sea level measurements into $1° \times 1°$ maps of sea level at 10-day intervals as described by Tapley *et al.* (1994a). If sea level at a given latitude (φ) and longitude (λ) map location is represented by $h(\varphi, \lambda, t)$, where t ranges over each T/P cycle, then the EOF representation of sea level can be written as

$$h(\phi, \lambda, t) = \sum_{i=1}^{N} S_i(\phi, \lambda) a_i(t),$$

where $S_i(\phi, \lambda)$ is the ith spatial EOF and $a_i(t)$ is its temporal history. These are determined from the eigenmodes of the spatial–temporal covariance matrix of the T/P sea level grids; thus each mode attempts to describe as much of the sea level variance as possible. Figure 6.13 (see color plate) shows the four leading (highest variance) EOF modes for sea level and SST (Reynolds and Smith, 1994) over the time frame of the T/P mission, where the temporal history of these modes has been scaled to represent their contribution to global mean sea level. Note that the first two "ENSO modes" describe most of the 1997–1998. ENSO event in global mean sea level. Mode 1 began rising at the beginning of 1997, peaked in early 1998 at 8 mm, and then fell back to zero by the end of 1998. Mode 2 began rising at the beginning of 1998, peaking at 11 mm near the end of 1998. Both of these modes show large signals in the tropics, as expected, but also large extratropical signals, especially in the Southern Ocean. Peterson and White (1998) have seen similar ENSO-related extratropical signals in sea surface temperature, which they attribute to atmospheric coupling between the two regions. Also note the large signal in the southwestern Indian Ocean between 5 and 10°S, which has been described by Chambers *et al.* (1998) and is correlated with ENSO in the Pacific. The EOF analysis establishes that most of the observed variations in global mean sea level are related to the ENSO phenomena.

Leuliette and Wahr (1999) have taken this idea one step further by conducting a coupled pattern analysis (Bretherton *et al.,* 1992; Wallace *et al.,* 1992) of the variations in mean sea level and SST (basically an EOF analysis of the cross covariance matrix of sea level and SST). This technique allows a more rigorous method of identifying the common EOF mode causing the changes in mean sea level, although the conclusions are basically the same as computing EOFs of the individual fields. Thus, they have also concluded that most of the long-term sea level change signal observed by T/P is being caused by changes in sea surface temperature related to the ENSO phenomena.

Clearly it would be desirable to remove the ENSO signals from the T/P sea level record so that we could begin to search for smaller signals, such as those related to climate change. It would be tempting to use the leading EOF modes (the ENSO modes) of sea level and SST in such a computation. However, these results must be interpreted carefully, as there is evidence that ENSO and climate change are not unrelated processes (Trenberth and Hoar, 1996), and thus removing the leading EOFs might also remove a climate signal. In addition, the EOF technique by no means guarantees that each mode will correspond to only a single process. In fact, processes that have spatial similarities will tend to coalesce into a single mode. There are many other statistical techniques that can be explored, so this should be considered an area of active research as of this writing.

6.5   THE FUTURE

6.5.1   Contribution to Climate Change Research

As mentioned in the first section, we are primarily interested in determining whether the rate of mean sea level is accelerating in response to global climate change, as suggested by the Intergovernmental Panel on Climate Change (IPCC) in their recent report (Houghton *et al.,* 1996). For the 20th century overall, no statistically significant acceleration of sea level has been found in tide gauge data (Douglas, 1992; Chapter 3; Woodworth, 1990), but tide gauge data do not possess the ability to detect a global acceleration in less than many decades. One strategy is to use the tide gauge estimate of the historical rate of 1.8 (± 0.1) mm/yr (Douglas, 1991) as the background rate, determine independently a rate from recent altimetric data, and then ask whether the altimetric rate is significantly different from the background rate. Although we will describe the present altimetric estimate of the recent sea level changes, we are mainly concerned with a careful error analysis of this approach in order to know what is required to make this calculation as sensitive as possible, and in particular how long of a time series will be required from satellite altimetry to detect climate signals. In the near future, however, we believe that it may be possible to quantitatively test the IPCC projections using this method. It will be much better when the acceleration of the rate can be measured directly using satellite altimetry, but this will require several decades of measurements.

We have shown that the 20-mm rise and fall of global mean sea level during 1997–1998 is almost certainly related to the ENSO phenomena. The presence of ENSO-scale variability in global mean sea level means that a longer time series will be needed to reduce this variability through averaging to reveal the smaller climate signals, unless reliable techniques can be developed to remove the ENSO signal from the data. To assess the impact of such variability, Nerem *et al.* (1999) developed two simulated long time series of global mean sea level variations, both containing ENSO variability. The first was an ~100-year-long (1882–1998) time series of the Southern Oscillation Index (SOI), which was linearly regressed against T/P mean sea level over 1993–1998 to determine the proper regression coefficients. The regression coefficients were then used to scale the SOI and simulate an ~100-year-long time series of global mean sea level. A similar technique was applied to a set of reconstructed global mean SST anomalies (Smith *et al.,* 1996) covering 1950–1998. In addition, the rate and acceleration of each of these time series were set to zero. Note that these estimates do not include errors in the tide gauge calibrations, but as mentioned previously, these errors should be significantly smaller than those caused by ENSO variability if geodetic monitoring is performed at each gauge.

Each of these time series was then used to perform two simulations. In the first simulation, a 2 mm/yr secular sea level change was added to the time series, and a series of Monte Carlo solutions was performed to determine the error in the estimated sea level rate by a least-squares solution using varying data spans with a random midpoint time. The result of these simulations is a plot of the accuracy of the sea level rise estimate versus the length of the data span of altimeter data employed (see Nerem *et al.,* 1999). Both simulations (using SOI or SST simulated sea level) suggest that 10 years of T/P class altimetry will be required to determine the rate of mean sea level change to an accuracy of 0.5 mm/yr. These simulations were repeated by adding an acceleration of mean sea level of 0.06 mm/yr^2 and estimating both the rate and the acceleration of sea level change. The SST results are invalid after about 20 years because the time series is of insufficient length to test longer data spans. However, the SOI results suggest that ~30 years of T/P class altimetry will be needed to detect an acceleration of mean sea level to an accuracy of 0.02 mm/yr^2. It should be noted that these results assume that the ENSO variability cannot be removed by other techniques; the situation could be improved considerably if this were so.

In summary, we should be able to detect a difference from the Douglas (1991) tide gauge–determined rate with a decade of T/P-class altimeter data, which will provide a limited test of the IPCC predictions. A more meaningful test, where the acceleration of mean sea level change is measured, will require at least 2–3 decades of similar data, unless decadal variability in global mean sea level is a significant contaminant to the computation. There is currently no way to assess the level of decadal variability in time series of global mean sea level computed using satellite altimetry. While two decades is a long time, it is much shorter than would be required using only the tide gauges.

If a sea level time series of several decades in length will be needed to test different sea level change predictions from climate change models, such a time series cannot be provided by a single satellite mission such as T/P; thus measurements from multiple missions will be needed to assemble a time series of sufficient length. A single 10-day averaged global mean sea level measurement from T/P has a precision of roughly 3–5 mm; thus we would like to link measurements from future missions to T/P with roughly the same accuracy.

Historically, this has proven to be a difficult task, as each mission has a unique set of time-varying measurement errors. Guman (1997) presented an analysis combining measurement from Geosat, ERS-1, and T/P. Because the ERS-1 and T/P missions significantly overlapped in time, establishing the relative bias between the two sets of measurements can be obtained through a straightforward differencing of their measurements at crossover points. However, the multiyear gap between the end of the Geosat mission and the beginning of the ERS-1 and T/P missions is more difficult to overcome. Tide gauge measurements spanning these missions provide the best basis for linking the sea level time series provided by these missions. However, the errors in this approach will be significant because of the unknown vertical movement of the tide gauges, and because of their poor geographic sampling of the altimeter measurement errors. This is particularly true for Geosat, which had much larger measurement and orbit errors than T/P. Guman (1997) implements a novel method for correcting the Geosat measurements using tide gauge comparisons; however, the combined Geosat/ERS-1/T/P sea level time series only loosely constrained the sea level change over the intervening decade.

6.5.2   The Future of Satellite Altimetry

The first missions likely to be used to extend the T/P measurement record are Jason-1 and Envisat, scheduled for launch in late 2000 and mid-2001, respectively. However, Envisat is in a sun-synchronous orbit, and thus tidal aliasing (Parke *et al.*, 1987) may be a significant error source since errors in the model of the solar tides will alias to zero frequency, potentially contaminating measurements of global mean sea level. If T/P is still operating when these spacecraft are launched, then computing the relative biases between these missions at a given time epoch will be relatively straightforward, and changes in these biases can be monitored using the tide gauges as discussed earlier. Indeed, it is currently planned to fly T/P and Jason-1 in tandem (in the same orbit separated in time by only a few minutes) for a short time to conduct a detailed study of the measurement differences. However, if the instruments on board T/P fail before the launch of these missions (there is enough propellant to maintain the orbit, but some of the satellite instruments have already exceeded their design lifetime by a factor of two!), it may be necessary to use tide gauges to bridge the measurement gap (Geosat Follow-On and ERS-2 may also be useful for bridging the gap, despite their larger errors relative to T/P). Envisat and Jason-1 will provide measurements of comparable accuracy to T/P, so the task of linking these measurements together will be much easier than encountered by Guman (1997) with Geosat. In this case, the largest error source is likely to be the vertical movement of the tide gauges. For a 1-year measurement gap, each tide gauge could be expected to move between 1 and 10 mm (due to postglacial rebound) if gauges in areas of known tectonic activity or subsidence are disregarded. For this reason, efforts to monitor the vertical motion of several tide gauges in the Pacific using the Global Positioning System have recently been initiated (Merrifield and Bevis, personal communication). A GPS-derived height time series of several years in length can determine the rate of vertical land motion to better than 1 mm/yr. Thus, while the problem will require careful study, it is likely that the T/P sea level time series can be extended in a seamless fashion using Envisat and/or Jason-1 measurements (the latter being preferable because they will fly in identical orbits). One caveat on this conclusion is the maintenance of the reference frame within which these measurements are made. As mentioned earlier, great care will have to be taken to ensure that the reference frame definition is maintained over several decades since the space geodetic technologies that are used to define the frame (and that also track the altimeter satellites) are undergoing rapid change as technology improves.

The first 6 years of T/P data have demonstrated that very precise measurements of global mean sea level can be made using satellite altimeters. In addition, the T/P results have shown that global mean sea level contains significant ENSO variability, which can be regarded as an error source for determining the long-term rate of sea level change, but is also of significant scientific interest in itself. We have shown that a series of T/P-Jason class satellite altimeter missions should be able to measure the long-term rate of sea level rise with about a decade of measurements, provided the instruments are precisely monitored using GPS-positioned tide gauges. Measuring the acceleration of sea level rise—a more important quantity for corroborating the climate models—will require 2–3 decades of measurements. We believe these objectives can be achieved given the current performance of T/P and the tide gauge calibration technique. Uncertainties in this assessment include (1) the unknown effects of decadal variability in global mean sea level,

(2) the possibility of gaps in the time series due to satellite failure, and (3) the successful implementation of GPS monitoring at each of the tide gauges. Despite these uncertainties, satellite altimetry is already defining a new paradigm in studies of sea level change.

Acknowledgments

We thank Eric Leuliette for producing some of the plots used in this paper, and Don Chambers for participating in many fruitful discussions regarding these results. The altimeter data base used in this study was developed by researchers at the Center for Space Research under the direction of Byron Tapley. This work was supported by a NASA Jason-1 Project Science Investigation.

REFERENCES

Anzenhofer, M., and Gruber, T. (1998). Fully reprocessed ERS-1 altimeter data from 1992 to 1995: Feasibility of the detection of long-term sea level change. *J. Geophys. Res.* **103** (CC4), 8089–8112.

Barnett, T. P. (1984). The estimation of "global" sea level change: A problem of uniqueness. *J. Geophys. Res.* **89** (C5), 7980–7988.

Bertiger, W. I., Bar-Sever, Y. E., Christensen, E. J., Davis, E. S., Guinn, J. R., Haines, B. J., Ibanez-Meier, R. W., Jee, J. R., Lichten, S. M., Melbourne, W. G., Muellerschoen, R. J., Munson, T. N., Vigue, Y., Wu, S. C., Yunck, T. P., Schutz, B. E., Abusali, P. A. M., Rim, H. J., Watkins, M. M., and Willis, P. (1994). GPS precise tracking of TOPEX/POSEIDON: Results and implications. *J. Geophys. Res.* **99** (C12), 24,449–24,464.

Birkett, C. M. (1995). The contribution of TOPEX/POSEIDON to the global monitoring of climatically sensitive lakes. *J. Geophys. Res.* **100** (C12), 25,179–25,204.

Birkett, C. M. (1998). Contribution of the TOPEX NASA radar altimeter to the global monitoring of large rivers and wetlands. *Water Resour. Res.* **34** (5), 1223–1240.

Bloomfield, P. (1976). *Fourier Analysis of Time Series: An Introduction.* Wiley, New York. 1976.

Born, G. H., Tapley, B. D., Ries, J. C., and Stewart, R. H. (1986). Accurate measurement of mean sea level changes by altimetric satellites. *J. Geophys. Res.* **91** (C10), 11,775–11,782.

Bretherton, C. S., Smith, C., and Wallace, J. M. (1992). An intercomparison of methods for finding coupled patterns in climate data. *J. Climate* **5,** 541–560.

Callahan, P. S., Hancock, D. W., and Hayne, G. S. (1994). New sigma 0 calibration for the TOPEX altimeter, *TOPEX/POSEIDON Research News* **3,** 28–32.

Cane, M. A., Kaplan, A., Miller, R. N., Tang, B., Hackert, E. C., and Busalacchi, A. J. (1996). Mapping tropical Pacific sea level: Data assimilation via a reduced state space Kalman filter. *J. Geophys. Res.* **101** (C10), 22,599–22,618.

Cazenave, A., Bonnefond, P., Dominh, K., and Schaeffer, P. (1997). Caspian sea level from TOPEX-Poseidon altimetry: Level now falling. *Geophys. Res. Lett.* **24** (8), 881–884.

Cazenave, A., Dominh, K., Gennero, M. C., and Ferret, B. (1998). Global mean sea level changes observed by Topex-Poseidon and ERS-1. *Phys. Chem. Earth* **23** (9–10), 1069–1075.

Cazenave, A., Dominh, K., Ponchaut, F., Soudarin, L., Cretaux, J.-F., and Le Provost, C. (1999). Sea level changes from Topex-Poseidon altimetry and tide gauge, and vertical crustal motions from DORIS. *Geophys. Res. Lett.* **26** (14), 2077–2080.

Chambers, D. P., Ries, J. C., Shum, C. K., and Tapley, B. D. (1998). On the use of tide gauges to determine altimeter drift. *J. Geophys. Res.* **103** (CC6), 12,885–12,890.

Chelton, D. B., and Schlax, M. G. (1996). Global observations of oceanic Rossby waves. *Science* **272**, 234–238.

Chen, J. L., Wilson, C. R., Chambers, D. P., Nerem, R. S., and Tapley, B. D. (1998). Seasonal global water mass budget and mean sea level variations. *Geophys. Res. Lett.* **25** (19), 3555–3558.

Chen, J. L., Wilson, C. R., Eanes, R. J., and Nerem, R. S. (1999). Geophysical interpretation of observed geocenter variations. *J. Geophys. Res.* **104** (B2), 2683–2690.

Cheney, R. E., Marsh, J. G., and Beckley, B. D. (1983). Global mesoscale variability from collinear tracks of Seasat altimeter data. *J. Geophys. Res.* **88**, 4343–4354.

Cheney, R., Miller, L., Agreen, R., Doyle, N., and Lillibridge, J. (1994). TOPEX/POSEIDON: The 2-cm solution. *J. Geophys. Res.* **99** (C12), 24,555–24,564.

Christensen, E. J., Haines, B. J., Keihm, S. J., Morris, C. S., Norman, R. A., Purcell, G. H., Williams, B. G., Wilson, B. D., Born, G. H., Parke, M. E., Gill, S. K., Shum, C. K., Tapley, B. D., Kolenkiewicz, R., and Nerem, R. S. (1994). Calibration of TOPEX/POSEIDON at Platform Harvest. *J. Geophys. Res.* **99** (C12), 24,465–24,486.

Dorandeu, J., and Le Traon, P. Y. (1999). Effects of global mean atmospheric pressure variations on mean sea level changes from TOPEX/Poseidon. *J. Atmos. Oceanic Tech.* **16** (9), 1279.

Douglas, B. C., Cheney, R. E., and Agreen, R. W. (1983). Eddy energy of the northwest Atlantic and Gulf of Mexico determined from GEOS 3 altimetry. *J. Geophys. Res.* **88** (14).

Douglas, B. C. (1991). Global sea level rise. *J. Geophys. Res.* **96** (C4), 6981–6992.

Douglas, B. C. (1992). Global sea level acceleration. *J. Geophys. Res.* **97** (C8), 12,699–12,706.

Douglas, B. C. (1995). Global sea level change: Determination and interpretation. *Rev. Geophys.* **33**, 1425–1432.

Fu, L.-L., and Glazman, R. (1991). The effect of the degree of wave development on the sea state bias in radar altimetry measurement. *J. Geophys. Res.* **96** (C1), 829–834.

Fu, L.-L., and Pihos, G. (1994). Determining the response of sea level to atmospheric pressure forcing using TOPEX/POSEIDON data. *J. Geophys. Res.* **99** (C12), 24,633–24,642.

Fu, L.-L., Christensen, E. J., Yamarone, Jr., C. A., Lefebvre, M., Menard, Y. M. e. n., Dorrer, M., and Escudier, P. (1994). TOPEX/POSEIDON mission overview. *J. Geophys. Res.* **99** (C12), 24,369–24,382.

Gaspar, P., Ogor, F., Le Traon, P.-Y., and Zanife, O.-Z. (1994). Estimating the sea state bias of the TOPEX and POSEIDON altimeters from crossover differences. *J. Geophys. Res.* **99** (C12), 24,981–24,994.

Groger, M., and Plag, H.-P. (1993). Estimations of a global sea level trend: Limitations from the structure of the PSMSL global sea level data set. *Global and Planetary Change* **8**, 161–179.

Guman, M. D. (1997). *Determination of Global Mean Sea Level Variations Using Multi-satellite Altimetry.* Ph.D. thesis, University of Texas, Austin.

Hayne, G. S., Hancock, D. W., and Purdy, C. L. (1994). TOPEX altimeter range stability estimates from calibration mode data. *TOPEX/POSEIDON Research News* **3**, 18–22.

Hendricks, J. R., Leben, R. R., Born, G. H., and Koblinsky, C. J. (1996). Empirical orthogonal function analysis of global TOPEX/POSEIDON altimeter data and implications for detection of global sea level rise. *J. Geophys. Res.* **101** (C6), 14,131–14,146.

Houghton, J. T., Meira Filho, L. G., Callander, B. A., Harris, N., Kattenberg, A., and Maskell, K. (1996). *Climate Change 1995.* Cambridge Univ. Press, Cambridge.

Imel, D. A. (1994). Evaluation of the TOPEX/POSEIDON dual-frequency ionosphere correction. *J. Geophys. Res.* **99** (C12), 24,895–24,906.

Kaula, W. M. (1970). *The Terrestrial Environment: Solid Earth and Ocean Physics,* Williamstown Report, NASA.

Kaula, W. M., Schubert, G., and Lingenfelter, R. E. (1974). Apollo Laser Altimetry and Inferences as to Lunar Structure, *Geochim. Cosmoschim. Acta Suppl.* **5**, 3049–3058.

Keihm, S., Zlotnicki, V., Ruf, C., and Haines, B. (1998). *TMR Drift and Scale Error Assessment.* Draft Report, Jet Propulsion Laboratory, 1–11.

Lame, D. B., and Born, G. H. (1982). SEASAT measurement system evaluation: Achievements and limitations, *J. Geophys. Res.* **87** (C5), 3175–3178.

Larnicol, G., Le Traon, P.-Y., Ayoub, N., and De Mey, P. (1995). Mean sea level and surface circulation variability of the Mediterranean Sea from 2 years of TOPEX/POSEIDON altimetry. *J. Geophys. Res.* **100** (C12), 25,163–25,178.

Le Traon, P. Y., and Gauzelin, P. (1997). Response of the Mediterranean mean sea level to atmospheric pressure forcing, *J. Geophys. Res.* **102** (C1), 973–984.

Leuliette, E. W., and Wahr, J. M. (1999). Coupled pattern analysis of sea surface temperature and TOPEX/POSEIDON sea surface height. *J. Phys. Oceanography* **29**, 599–611.

Marshall, J. A., and Luthcke, S. B. (1994). Modeling radiation forces acting on TOPEX/ POSEIDON for precision orbit determination, *J. Spacecraft and Rockets,* **31** (1), 99–105.

Marshall, J. A., Zelensky, N. P., Klosko, S. M., Chinn, D. S., Luthcke, S. B., Rachlin, K. E., and Williamson, R. G. (1995). The temporal and spatial characteristics of TOPEX/POSEIDON radial orbit error. *J. Geophys. Res.* **100** (C12), 25,331–25,352.

Maul, G. A., and Martin, D. M. (1993). Sea level rise at Key West, Florida, 1846–1992: America's longest instrument record? *Geophys. Res. Lett.* **20** (18), 1955–1958.

McGoogan, J. T., Miller, L. S., Brown, G. S., and Hayne, G. S. (1974). The S-193 Radar Altimeter Experiment, *IEEE Trans.* (June).

Melbourne, W. G., Davis, E. S., Yunck, T. P., and Tapley, B. D. (1994). The GPS flight experiment on TOPEX/Poseidon. *Geophys. Res. Lett.* **21** (19), 2171–2174.

Minster, J.-F., Brossier, C., and Rogel, P. (1995). Variation of the mean sea level from TOPEX/ POSEIDON data. *J. Geophys. Res.* **100** (C12), 25,153–25,162.

Minster, J. F., Cazenave, A., Serafini, Y. V., Mercier, F., Gennero, M. C., and Rogel, P. (1999). Annual cycle in mean sea level from Topex-Poseidon and ERS-1: Inference on the global hydrological cycle. *Global and Planetary Change* **20,** 57–66.

Mitchum, G. T. (1994). Comparison of TOPEX sea surface heights and tide gauge sea levels. *J. Geophys. Res.* **99** (C12), 24,541–24,554.

Mitchum, G. T. (1998). Monitoring the stability of satellite altimeters with tide gauges. *J. Atmos. and Oceanic Tech.* **15,** 721–730.

Mitchum, G., Kilonsky, B., and Miyamoto, B. (1994). Methods for maintaining a stable datum in a sea level monitoring system. In *OCEANS 94 OSATES,* IEEE Oceanic Engineering Society, Brest, France.

Molines, J. M., Le Provost, C., Lyard, F., Ray, R. D., Shum, C. K., and Eanes, R. J. (1994). Tidal corrections in the TOPEX/POSEIDON geophysical data records. *J. Geophys. Res.* **99** (C12), 24,749–24,760.

Moore, P., Carnochan, S., and Walmsley, R. J. (1999). Stability of ERS Altimetry during the Tandem Mission. *Geophys. Res. Lett.* **26** (3), 373–376.

Morris, C. S., and Gill, S. K. (1994). Evaluation of the TOPEX/POSEIDON altimeter system over the Great Lakes. *J. Geophys. Res.* **99** (C12), 24,527–24,540.

Nerem, R. S. (1995a). Global mean sea level variations from TOPEX/POSEIDON altimeter data. *Science* **268,** 708–710.

Nerem, R. S. (1995b). Measuring global mean sea level variations using TOPEX/POSEIDON altimeter data. *J. Geophys. Res.* **100** (C12), 25,135–25,152.

Nerem, R. S. (1997). Global mean sea level change: Correction. *Science* **275,** 1053.

Nerem, R. S. (1999). Sea level change. In *Satellite Altimetry* (L. Fu, and A. Cazenave eds.), Academic Press, San Diego.

Nerem, R. S., Lerch, F. J., Marshall, J. A., Pavlis, E. C., Putney, B. H., Tapley, B. D., Eanes, R. J., Ries, J. C., Schutz, B. E., Shum, C. K., Watkins, M. M., Klosko, S. M., Chan, J. C., Luthcke, S. B., Patel, G. B., Pavlis, N. K., Williamson, R. G., Rapp, R. H., Biancale, R., and Nouel, F. (1994). Gravity model development for TOPEX/POSEIDON: Joint Gravity Models 1 and 2. *J. Geophys. Res.* **99** (C12), 24,421–24,448.

Nerem, R. S., Haines, B. J., Hendricks, J., Minster, J. F., Mitchum, G. T., and White, W. B. (1997a). Improved determination of global mean sea level variations using TOPEX/ POSEIDON altimeter data. *Geophys. Res. Lett.* **24** (11), 1331–1334.

Nerem, R. S., Rachlin, K. E., and Beckley, B. D. (1997b). Characterization of global mean sea level variations observed by TOPEX/POSEIDON using empirical orthogonal functions. *Surveys in Geophys.* **18,** 293–302.

Nerem, R. S., Eanes, R. J., Ries, J. C., and Mitchum, G. T. (1998). The use of a precise reference frame for sea level change studies. In *Integrated Global Geodetic Observing System,* International Association of Geodesy, Munich.

Nerem, R. S., Chambers, D. P., Leuliette, E. W., Mitchum, G. T., and Giese, B. S. (2000). Variations in global mean sea level associated with the 1997–98 ENSO event: Implications for measuring long term sea level change. *Geophys. Res. Lett.* in review.

Nouel, F. N. e. l., Berthias, J. P., Deleuze, M., Guitart, A., Laudet, P., Piuzzi, A., Pradines, D., Valorge, C., Dejoie, C., Susini, M. F., and Taburiau, D. (1994). Precise Centre National d'Études Spatiales orbits for TOPEX/POSEIDON: Is reaching 2 cm still a challenge? *J. Geophys. Res.* **99** (C12), 24,405–24,420.

Parke, M. E., Stewart, R. H., Farless, D. L., and Cartwright, D. E. (1987). On the choice of orbits for an altimetric satellite to study ocean circulation and tides. *J. Geophys. Res.* **92,** 11693–11707.

Peterson, R. G., and White, W. B. (1998). Slow oceanic teleconnections linking the Anarctic Circumpolar Wave with the tropical El Niño-Southern Oscillation. *J. Geophy. Res.* **103** (C11), 24,573–24,583.

Preisendorfer, R. W. (1988). *Principal Component Analysis in Meteorology and Oceanography.* Elsevier.

Pugh, D. (1987). *Tides, Surges, and Mean Sea Level.* Wiley, New York.

Raofi, B. (1998). *Ocean Response to Atmospheric Pressure Loading: The Inverted Barometer Correction for Altimetric Measurements.* University of Texas, Austin.

Reynolds, R. W., and Smith, T. S. (1994). Improved global sea surface temperature analysis. *J. Climate* **7,** 929–948.

Ruf, C. S., Keihm, S. J., Subramanya, B., and Janssen, M. A. (1994). TOPEX/POSEIDON microwave radiometer performance and in-flight calibration. *J. Geophys. Res.* **99** (C12), 24,915–24,926.

Russell, G. L., Miller, J. R., and Rind, D. (1995). A coupled atmosphere-ocean model for transient climate change. *Atmosphere-Ocean* **33** (4), 683–730.

Russell, G. L., Miller, J. R., Rind, D., Ruedy, R. A., Schmidt, G. A., and Sheth, S. (2000). Climate simulations by the GISS Atmosphere-Ocean Model: 1950 to 2099. *Geophys. Res. Lett.* in review.

Schutz, B. E., Tapley, B. D., Abusali, P. A. M., and Rim, H. J. (1994). Dynamic orbit determination using GPS measurements from TOPEX/Poseidon. *Geophys. Res. Lett.* **21**(19), 2179–2182.

Shum, C. K., Woodworth, P. L., Andersen, O. B., Egbert, G. D., Francis, O., King, C., Klosko, S. M., Le Provost, C., Li, X., Molines, J.-M., Parke, M. E., Ray, R. D., Schlax, M. G., Stammer, D., Tierney, C. C., Vincent, P., and Wunsch, C. I. (1997). Accuracy assessment of recent tide models. *J. Geophys. Res.* **102** (C11), 25173–25194.

Smith, T. M., Reynolds, R. W., Livezey, R. E., and Stokes, D. C. (1996). Reconstruction of historical sea surface temperatures using empirical orthogonal functions. *J. Climate* **9,** 1403–1420.

Stammer, D., Tokmakian, R., Semtner, A., and Wunsch, C. (1996). How well does a 1/4 deg. global circulation model simulate large-scale oceanic observations? *J. Geophys. Res.* **101** (C11), 25,779–25,812.

Stanley, H. R. (1979). The Geos 3 Project. *J. Geophys. Res.* **84** (B8), 3779–3783.

Stewart, R. (1985). *Methods of Satellite Oceanography.* Univ. California Press, Berkeley and Los Angeles.

Tapley, B. D., Chambers, D. P., Shum, C. K., Eanes, R. J., Ries, J. C., and Stewart, R. H. (1994a). Accuracy assessment of the large-scale dynamic ocean topography from TOPEX/POSEIDON altimetry. *J. Geophys. Res.* **99** (C12), 24,605–24,618.

Tapley, B. D., Ries, J. C., Davis, G. W., Eanes, R. J., Schutz, B. E., Shum, C. K., Watkins, M. M., Marshall, J. A., Nerem, R. S., Putney, B. H., Klosko, S. M., Luthcke, S. B., Pavlis, D., Williamson, R. G., and Zelensky, N. P. (1994b). Precision orbit determination for TOPEX/POSEIDON. *J. Geophys. Res.* **99** (C12), 24,383–24,404.

Tapley, B. D., Shum, C. K., Ries, J. C., Suter, R., and Schutz, B. E. (1992). Global mean sea level using the Geosat altimeter. In *Sea Level Changes: Determination and Effects,* American Geophysical Union, IUGG.

Tapley, B. D., Watkins, M. M., Ries, J. C., Davis, G. W., Eanes, R. J., Poole, S. R., Rim, H. J., Schutz, B. E., Shum, C. K., Nerem, R. S., Lerch, F. J., Marshall, J. A., Klosko, S. M., Pavlis, N. K., and Williamson, R. G. (1996). The Joint Gravity Model 3. *J. Geophys. Res.* **101** (B12), 28,029–28,050.

Trenberth, K. E., and Hoar, T. J. (1996). The 1990–1995 El Niño-Southern Oscillation event: Longest on record. *Geophys. Res. Lett.* **23** (1), 57–60.

Wagner, C. A., and Cheney, R. E. (1992). Global sea level change from satellite altimetry. *J. Geophys. Res.* **97** (C10), 15,607–15,615.

Wallace, J. M., Smith, C., and Bretherton, C. S. (1992). Singular value decomposition of wintertime sea surface temperature and 500-mb height anomalies. *J. Climate* **5**, 561–576.

Woodworth, P. L. (1990). A search for acceleration in records of european mean sea level. *Int. J. Climatology* **10**, 129–143.

Wyrtki, K., and Mitchum, G. (1990). Interannual differences of GEOSAT altimeter heights and sea level: The importance of a datum. *J. Geophys. Res.* **95** (C3), 2969–2975.

Yunck, T. P., Bertiger, W. I., Wu, S. C., Bar-Server, Y. E., Christensen, E. J., Haines, B. J., Lichten, S. M., Muellerschoen, R. J., Vigue, Y., and Willis, P. (1994). First assessment of GPS-based reduced dynamic orbit determination on TOPEX/Poseidon. *Geophys. Res. Lett.* **21**(7), 541–544.

Chapter 7 | Decadal Variability of Sea Level

W. Sturges
B. G. Hong

7.1 INTRODUCTION

It is difficult to address the sea level rise problem using tide gauge data, because water level records reflect many processes in addition to a secular trend. We have seen in Chapter 3 that tide gauge records appear very "noisy" due to seasonal-to-interannual and longer variations of water level. The purpose of this chapter is to use the longest tide gauge records, plus wind data and the appropriate physics, to see if there is hope for improving the signal-to-noise ratio. The results are very positive; interannual and longer variations of sea level on the U.S. east coast can be explained and modeled.

Some natural processes, such as the astronomical tides, have known, simple periods associated with them. It is surprising, therefore, to find energetic frequency bands (i.e., certain periods) in which the conspicuous variability is associated with an almost random forcing. The term "decadal" in the title of this chapter (and elsewhere) is used loosely. While there certainly *is* power at periods near a decade, power is actually spread over a wide frequency band. In normal usage the term implies "longer than a year and extending out beyond 10 years." If one should ask, How much longer than a decade? we must answer with chagrin, We don't know! The periods of "decadal" sea level signals are longer than we can treat rigorously; existing measurement records are not long enough to enable treating them rigorously in a statistical sense.

Examples of this long-period variability are shown in figures found in other chapters of this book, especially Chapter 3, or for example Woodworth (1990) or Douglas (1992). The decadal variability is not a new problem; Fig. 7.1 is patterned after the work of Hicks and Crosby (1974; or see, for example, Hicks *et al.*, 1983). These are now-classic examples of sea level variability at a few U.S. tide gauges. Data such as these are available at the Web sites of the National Ocean Service (www.nos.noaa.gov) and the Permanent Service for Mean Sea level (www.pol.ac.uk) and in the CD-ROM that accompanies this volume. These figures show how similar the variability can appear to be

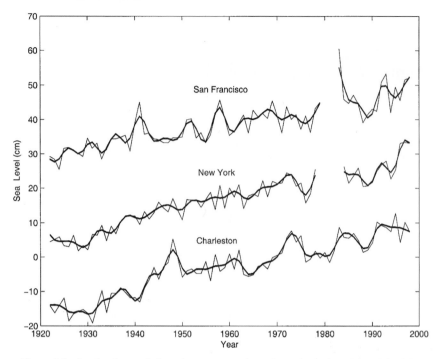

Figure 7.1   Sea level records from long-term stations along the coasts of the United States. The individual data points are the annual mean values; the solid line shows a low-passed version that retains variability at periods longer than 5 years but suppresses variability at periods shorter than 2.5 years.

along our coasts, although a sharp eye will quickly find particular events at one location that are not duplicated at others. But the records in Fig. 7.1 do have in common an overall increase of sea level. In Alaska, however, as in many high-latitude locations near glaciated regions, the long-term trend is clearly a fall of sea level. Geology, not oceanography, dominates at such locations.

Several important features in Fig. 7.1 deserve emphasis. First, while it is not apparent in the figures, there is an enormous amount of variability at higher frequencies that has been suppressed (or "filtered out"). Some details about the necessary signal processing are included in Appendix 7.1. Second, the noticeable bumps and wiggles are sometimes similar between New York and Charleston, as in the early 1970s, and sometimes not, as for example the large peak in the late 1940s at Charleston that is not found at New York. These features, perhaps to the surprise of many, are a result not of coastal processes but of the large-scale winds over the Atlantic Ocean.

The most important feature of Fig. 7.1 is that sea level rises over time at these mid-latitude stations—a phenomenon seen in nearly all long mid-latitude records. Typical values around the United States are approximately 2–4 mm/ yr, or in ordinary terms, ~1 foot per century. While progress has been made in determining and understanding the worldwide increase of sea level, we are only now beginning to understand the causes of the decadal fluctuations. This topic is addressed later in this chapter.

The rise of sea level and its variability along mid-latitude coasts are well known; Sturges *et al.* (1998) have shown a direct connection between the decadal variability along the coast and similar fluctuations in the open ocean. However, in the central North Atlantic, at the latitudes of the subtropical gyre, there is but one island tide gauge: Bermuda. It is well known, however, that at low frequencies the tide gauges at islands do a good job of representing the ocean on large scales. See, for example, Wunsch (1972) or Roemmich (1990). In contrast to the scarcity of sea level records in the broad oceans, along the U.S. east and west coasts there is a wealth of tide gauge data. The general features of these tide gauge records are a gradual rise of sea level imbedded in a background of low-frequency variability: see, for example, the papers of Barnett (1984), Douglas (1991, and Chapter 3 of this book), Woodworth (1990), and Wunsch (1992).

Our main interest here is sea level at the coast. However, it is important for the analysis that follows to recognize an important point: sea levels at mid-ocean tide gauges, such as Bermuda, or the equivalent variations derived from subsurface density data (Levitus, 1990) show variability much like that on the U.S. east coast: low-frequency variations having peak-to-peak amplitudes over 10 cm.

For examples of coastal sea level data, we show one each from the U.S. east and west coasts. The records at Charleston and San Francisco are some of the longest. Figure 7.2 shows the power spectra[1] at these stations in two ways: the first panel, Fig 7.2a, shows what most in this field would call the usual method, which uses a log-log presentation. In the second panel, 7.2b, we use the variance-preserving method, which shows to the eye where (in frequency) most of the power lies. It would appear from the first panel that the power in these records is concentrated in a few "lumps" of energy in certain frequency bands. Even though these records are long, we should not attach much physical significance to the power in these specific frequency bands. If we cut the records into halves and compute the spectra for each half, we find for both of these long records (as well as at most others) that the high spots in one half correspond to the low spots in the other. That is, the frequency distribution is seemingly random and varies enormously from decade to decade and from place to place.

---

[1] These spectra are derived from the records of monthly data. The power peak at a year (frequency 0.0833/month) is typical of coastal gauges and is consistent with observations offshore. The annual cycle at these latitudes results largely from the annual cycle of stored heat. The other spikes are at sub-multiples of 12 mo.: 6 mo, 4 mo, etc.

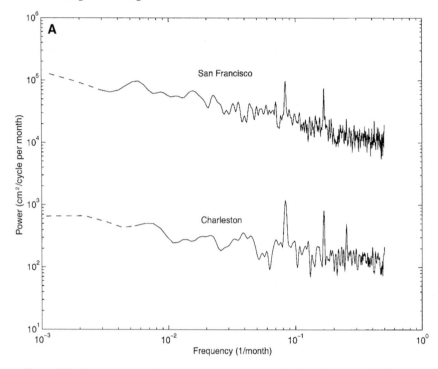

Figure 7.2 Power spectra of two long sea level records: San Francisco, California, and Charleston, South Carolina. Panel A shows the ordinary version, with log-log scales; Panel B shows the so-called variance-preserving form, in which the power in each frequency region is proportional to the area apparent to the eye. The smoothing is by 5 Hanning passes at the higher frequencies, decreasing toward lower frequencies to show the variance at the longest periods.

The stations that have the longest records usually show the long-term sea level rise most clearly. This is so because the fluctuations on the order of 10–20 cm peak-to-peak are typical of the decadal sea level variability. And this variability, of course, obscures changes in the underlying trend.

7.1.1 Magnitude of the Long-Term Rise of Sea Level

Suppose we wish to determine the slope of the sea level signals at New York, for example (see Fig. 7.1). It is a trivial task to make a least-squares, straight-line fit to any set of data. If all we want is a good estimate, a pencil and a ruler will do. Yet suppose for a hypothetical record that the variability from decadal signals caused sea level to be unusually low at the beginning of our record, such as in the late 1920s at New York or 1945 at San Francisco.

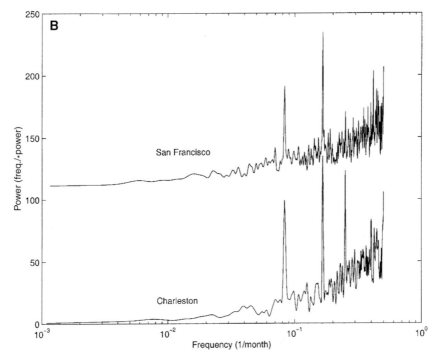

Figure 7.2   (*Continued*)

And by essentially random events, suppose that in our hypothetical record the sea level was unusually high at the end, such as near 1920 at New York or at 1940 at San Francisco. As discussed in detail in Chapter 3, such anomalies would lead to an apparent slope that would be much too great compared with the true long-term trend. With the benefit of a considerably longer record, we are able to detect such anomalies. We only know that the computed trend is too great because, in our hypothetical example, we can detect that the random fluctuations are low at the beginning and high at the end. We can detect this only if we have an "adequately long" record in some sense. In general, of course, we do not have such insight.

A person trained in signal processing will tell us that it is also possible to filter these decadal signals—to remove the undesired variations—to reveal the underlying trend. But two problems arise. First, what periods (or frequencies) shall we remove? This turns out to a most difficult choice—one not provided by the external parameters of the problem. And second, after an appropriate filtering operation, it is necessary in high-accuracy applications to discard long segments at the ends of the record.[2] After much filtering, therefore, little signal remains!

Figure 7.2 shows the spectra of two of the long-term tide gauges. The "decadal signals" causing the problems just described have their main power at periods longer than 100 months. (Although the absolute amount of power in the records is not large at these long periods, as can be seen in Fig. 7.2B (p. 169), when we examine the low-frequency motions for the sea level rise problem, this power becomes the dominant issue.) The power continues to increase out to the lowest frequencies; this tendency is called a "red spectrum," somewhat in analogy to red light appearing at the low-frequency end of the visible spectrum. It requires care to obtain a meaningful spectrum when the power is found at periods that are not short relative to the length of the record. Most standard methods designed to be used in electrical engineering, for example, are constructed with the implicit assumption that one can obtain a signal of any necessary duration. Such methods are not appropriate in most oceanographic or geophysical situations. To remove this low-frequency power by careful filtering would require that we discard on the order of 100–200 months of signal at each end of the record. Different filtering methods produce different results, but this is an essential feature of accurate analysis. In other words, one is faced with the necessity of discarding on the order of 200–400 months of signal if common filtering techniques are to be employed. This is a major fraction of the length of most records! For a more familiar example, observing a single tidal cycle of the semidiurnal tide would give us a nice observation of one cycle of something, but no knowledge about the fortnightly tides. A general rule of thumb in such analyses is that until we have seen roughly 10 cycles we do not have enough information to determine meaningful statistics about the process. As will be shown below, it appears that the seemingly random fluctuations in sea level data have periods at least as long as 20 years, so the rule of thumb is not satisfied.

7.1.2   Determining a Change in the Rate of Sea Level Rise

Now on to the most disturbing fact about the problems associated with decadal variability. This discussion has focused on the difficulty of obtaining the slope of the curves of Fig. 7.1. But our true task here is to study the effects (if any) of man-made changes in the rate of sea level rise. That is, we would like to put a reliable straight-line fit to the first part of the record, a straight-line fit to the last part of the record where presumably anthropogenic effects would be more pronounced, and determine the change of slope where the two lines meet. This is a challenge for two reasons. First, for the reasons described

---

[2] Some processing methods suggest "tricks", such as reflecting the signal back upon itself at the ends, that allow us to retain the full record length in the filtered result. These of course are equivalent to knowing the data before the record began and after it has ended. Such divine insight is desirable but difficult to achieve.

above, it is technically difficult. Second, because the low-frequency variability is as large as the "effect" we wish to document, different analysts will perceive different places in the record at which the slope seems to change. For example, the records at New York and Charleston suggest an apparent increase in the rate of rise beginning in the late 1920s. Is this the effect of man-made emissions, or just a random transient? Clearly, if such an effect is a real physical event, we would expect to find similar effects at all major tide gauges. To study this problem further, consider tide gauge data taken at San Francisco in more detail.

Figure 7.3 gives a closer look at the data portion prior to 1900. San Francisco has one of the few gauges where data of this duration are available. Sea level seems to be approximately stable if we examine the record from 1865 to 1900. However, if we chose to begin our analysis in the early 1880s, we would see a fall of sea level at the rate of 20 cm/100 yr for a few decades during the late 1800s! Therefore, if we believe, on the basis of a great quantity of other data, that there is a long-term general rise of sea level, we are forced to conclude that the period of sea level fall during the late 1800s at San Francisco is an example of a transient anomaly—perhaps a random event. If this feature is taken as an example of how large a random transient feature can be, we must

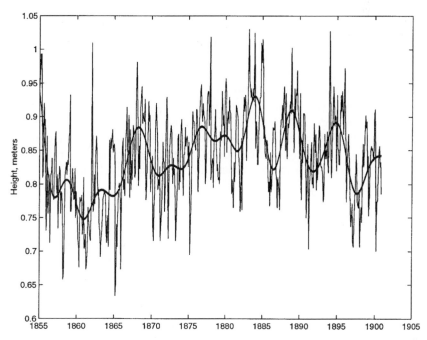

Figure 7.3 Sea level at San Francisco, extending back to the 1850s, from monthly mean values. This plot shows only the portion before 1900, to emphasize the interval of declining sea level after the early 1880s.

conclude that such events can have durations as long as 20 years. This is not a happy conclusion, but we ignore it at our peril. How can we be confident that the next such fluctuation will not be longer?

The appropriate scientific question becomes: What phenomena cause these low-frequency signals? Can we understand them well enough to remove them on the basis of physical principles? We show in a later section that the culprit is wind over the open ocean.

7.2   A CLOSER EXAMINATION OF SOME IMPORTANT DETAILS: LOW-FREQUENCY WAVES IN THE OCEAN

7.2.1   Observed Signals in the Atlantic Ocean

A meaningful question to consider is: How big is the interdecadal sea level signal in the open ocean, and why is the open-ocean signal relevant to coastal variability? Workers in other fields are often surprised by the nature of low-frequency, open-ocean waves. The idea that an ocean wave can have periods of many years is usually an amazing revelation. Yet this is clearly the case. Figure 7.4, an updated version of the original figure shown in Sturges and Hong (1995), shows a comparison between sea level at Bermuda computed from a numerical model and the tide gauge there. Because this comparison is strikingly good, we tend to believe that the model output is reliable to within a few centimeters at these long periods. The first point to notice is that these waves have amplitudes of nearly 20 cm peak-to-peak—the same as the decadal signals observed at the coast. These waves travel at a few centimeters per second (see Sturges *et al.,* 1998, for details as to the wave speeds), so the travel time across the Atlantic is roughly 5 years or more. These are forced waves, and the forcing is from the large-scale winds over the ocean. The important point here is that these long so-called Rossby waves travel only to the west. They strike the western side of the oceans, or the eastern coasts of the continent, creating a sharp distinction between sea level variations on east coasts and west coasts.

7.2.2   A Rossby Wave Model

Rossby waves, the dominant low-frequency waves of the ocean and the atmosphere, are described by Platzman (1968). The dominant restoring force for these waves arises not merely from the Coriolis force, which is simply an effect of the rotation of the earth, but from the curious fact that, for waves of such large, planetary scale, the Coriolis force changes with latitude. Almost all large signals in the ocean or weather systems in the atmosphere propagate via Rossby waves. In the ocean, the long, low-frequency waves travel to the west. In the atmosphere (weather systems for example), by contrast, they travel to the east.

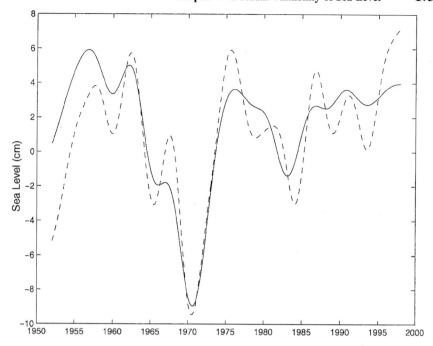

Figure 7.4   Comparison between observed sea level at Bermuda and a Rossby wave model calculation. Sea level at Bermuda is shown by the dashed curve; the model output is shown by the solid curve. The model is forced only by open ocean winds. This is an updated version of the similar figure shown by Sturges and Hong (1995).

The model of wind forcing that we have used for the results presented here is patterned after Meyers (1979) and Kessler (1990) and was used by Sturges and Hong (1995) and by Sturges *et al.* (1998). The model that produced the solid curve in Fig. 7.4 concentrates on forcing at periods longer than a few years, at which the ocean's response will be primarily from long, nondispersive, baroclinic Rossby waves. Price and Magaard (1986) found that waves at these long periods traveled within a few degrees of due westward. Mayer and Weisberg (1993) made an important contribution by showing that the frequency content of the observed wind forcing is very much like the frequency content of sea level—a most important clue. Levitus (1990) was one of the first to show solid evidence that the ocean undergoes such low-frequency variations. Greatbatch *et al.* (1991), Ezer *et al.* (1995), and others have shown that in the open ocean, numerical models can reproduce the observed oceanic variability. Our focus here of course is on showing that such results also apply directly at the coast.

The model we have used is intentionally simple. We separate the ocean's response into a series of what are called vertical modes. It is easy to envision the way in which vertical modes work. They are exactly like the shapes of the various harmonics one can make by plucking a violin string—or the spring on a screen door. For the violin string, the tension is the same everywhere, so we get uniform sine waves; in the ocean, the shapes are a little different because the vertical density stratification (analogous to the string tension) changes with depth; other than that the physical problem is the same. This gives the result that the low-frequency, large-scale wind-driven response for fluctuations of the sea surface (for the nth mode) is described by the product of a vertical structure function (whose shape is standard and need not concern us here) and a pressure function that is a function of space and time, as

$$\frac{\partial p_n}{\partial t} - \gamma_n \frac{\partial p_n}{\partial x} = B_n \text{ curl } (\tau/\rho_0 f) \tag{1}$$

where p is pressure; t is time; x is the distance eastward, y northward, and z upward; τ is wind stress; f is the Coriolis parameter; ρ_0 is a constant representative water density; and γ_n is the long Rossby wave speed for the nth vertical mode. The term B is an amplitude factor determined by the vertical structure function. Details are given in Sturges *et al.* (1998). The wave speed, γ, can be computed directly from the (known) density distribution in the vertical direction. One very great advantage of our model formulation is that everything in the model is either physics or observations; there are no "free parameters" to tune to make the results look right. The wind stress, τ, is obtained from an oceanic wind compilation, the Comprehensive Ocean-Atmosphere Data Set, or "COADS" archive (Slutz *et al.,* 1985).

The physical meaning of the terms in (1) is that the first term is the time rate of change of pressure at any point, which could be at the sea surface or deeper. The second term, the product of the wave speed, γ, times the local slope of the pressure surface, represents the rate at which the free waves travel, given no external forcing. The last term represents the forcing of the large-scale winds. The curl is the familiar mathematical operator; this term is often call the "Ekman pumping" because it represents the local convergence of any surface "Ekman transport" by the local winds. That is, if more surface transport is being driven into the region than is going out the other side, there must be a convergence; the only place it can go is down, forcing near-surface water deeper into the ocean. Such vertical velocities are extremely small (order of 10^{-5} cm/s) but of major dynamical importance in the ocean. Of course, this term can be positive or negative.

An important thing to notice here is that the model contains no frictional terms that can be conveniently "tuned" to get an answer we like. The wave speeds, γ, are computed a priori from the density field, as are the vertical structure functions. The wind stress is determined from historical ships' observations (available from the National Climatic Data Center at www.ncdc.noaa. gov). The model produces an output that mimics the variability of the sea surface of the central North Atlantic Ocean since the end of WWII; a simple movie of this output is presented on the CD-ROM that accompanies this volume.

Figure 7.4 has the mean trend, approximately 18 cm/century (the "rise of sea level" signal), removed, since there is a long-term rise of sea level even in mid-ocean. The important point, however, is that we know the signals at Bermuda are propagating to the west. We can be certain, therefore, that they will strike the western boundary of the ocean. Thus we are led to a very important conclusion: the decadal variability of sea level on the U.S. east coast is most likely the result of the impact of such waves.

7.2.3   Observed Signals in the Pacific Ocean

The fluctuations in the sea level record at Hawaii are not terribly different, qualitatively, from those at Bermuda. And the waves that travel past Hawaii are going to the west, just like those at Bermuda. They will never reach the U.S. mainland. The only waves sufficiently energetic to reach the U.S. west coast, by contrast, are those that are trapped at the equator and which (because of quite different physics) travel to the east. Large, low-frequency signals at near-equatorial latitudes, such as the famous ENSO (El Niño, Southern Oscillation) signals, are rather fast and travel along the equator from west to east. When they strike the continent, they then propagate poleward along the coast, in both hemispheres. See, for example, Clarke (1992) for details of the reflection at the boundary and Chelton and Davis (1982) for an insightful discussion of the processes by which waves propagate up the U.S. west coast. White (1977) made an interesting discovery about *annual* period waves that travel across the Pacific, but of course, the frequencies of interest for the sea level rise problem are at periods of many years. Other studies of wave propagation in the Pacific include those of Qui (1997) and Cummins *et al.* (1986).

7.3   THE UNDERLYING PHYSICS OF THE PROBLEM

We have only recently learned that the cause of the decadal signals on the east coast of continents is from wind forcing over the open ocean. This work is described by Hong *et al.* (2000). Other candidate forcing mechanisms such as variations in local winds along the coast or salinity variations caused by changes in river runoff effects have been studied and found to be unimportant for these sea level signals (Hong *et al.* 2000). Although the propagation of long Rossby waves across the ocean is a relatively simple process to model, the question of how to understand the passage of these waves across or through the Gulf Stream was the major problem attacked by Hong *et al.* (2000).

In general, we would expect that as a long Rossby wave passes through the Gulf Stream, the strong horizontal shear in the stream would destroy most of the identifiable wave characteristics. Yet it has long been known that at periods as long as a year the annual cycle of stored heat causes the ocean to go up and down at the coast just as out in the open ocean. In the case of the decadal signals, however, the waves clearly must propagate through the Gulf Stream to reach the southern U.S. east coast, and it was not understood how this could happen.

Figure 7.5, taken from Hong *et al.* (2000), shows a comparison between the observed sea level variability at a coastal station on the coast of Delaware and the computed sea level variability from the Rossby wave model. The mean slope has been removed. The importance of this result is that the low-frequency (interannual and longer) variability at the coast has been computed entirely from a model that uses only winds over the full width of the Atlantic Ocean as its forcing mechanism. Therefore the addition of variations in salinity, of river inputs, of atmospheric pressure variability, or of local winds along the coast can only serve to make small, minor improvements in the results shown in Fig. 7.5.

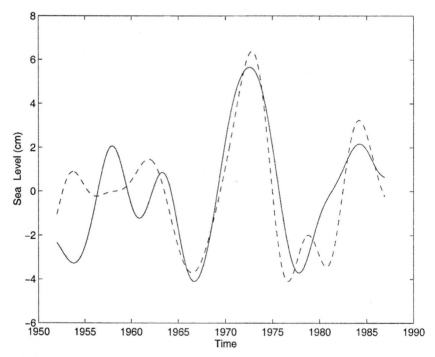

Figure 7.5 Comparison between coastal sea level and a Rossby wave model calculation. Sea level at Lewes, Delaware, on the U.S. east coast, is shown by the dashed curve; the model output is shown by the solid curve. The model is forced only by open ocean winds. Adapted from Hong *et al.* (2000).

It is worth emphasizing that the sea level signals of Fig. 7.5 are produced by the interactions between the winds and the ocean within the model of Hong *et al.* (2000). That is, they are an oceanic phenomenon and not in the wind field itself. To put it another way, the model *output* of Fig. 7.5 is in no way a simple linear modification of the input wind forcing signals. It represents the complex response of the ocean to a nearly random wind forcing. Although we understand this process enough to model it adequately, the underlying nature of this response is not thoroughly understood.

7.3.1   Correcting Observed Sea Level Rise for Wind-Induced Decadal Signals

At this point the next logical step should be obvious: using the ocean winds as inputs, we compute the model sea level signals and subtract the wind-forced part to determine the true "rise of sea level" at the coasts. Unfortunately, doing the calculation requires knowing the wind field. These long term winds are not available. During World Wars I and II, taking wind data over the oceans took a back seat to hiding convoys of ships from submarines. As a result, there are large gaps in the wind data set during the war years, making it much more difficult to calculate the wind-induced sea level signals. There are techniques for attempting to interpolate across gaps in a record, but one has little assurance that simple methods produce reliable results. No one knows, obviously, what the winds were, but certain statistical properties of the wind field can be assumed to remain unchanged from one time period to the next. This is clearly a research-level project, and the authors are attempting to complete the work even as this is written.

APPENDIX: SOME DETAILS OF THE SIGNAL PROCESSING OF TIDE GAUGE DATA

Figure 7.A.1 (upper panel) shows an example at Charleston, South Carolina, of what the signal looks like when we do not remove the tides or other high-frequency motions. Figure 7.A.1 (lower panel) shows these variations after the tides have been removed but the month-to-month variability has not been removed. The most noticeable variability is an irregular annual cycle. This mid-latitude annual variability has been shown to be caused by the stored heat in the upper hundred meters of the ocean, which is in turn associated with the annual solar cycle.

The variability in Fig. 7.A.1 (lower panel) however, is obviously very irregular. The reason for this, perhaps surprisingly, is that the strongest effect at periods of a few days is associated with longshore currents, associated in turn with local wind forcing. Figure 7.A.2 shows the signal in this so-called "wind band," or synoptic band. The variations shown in Fig. 7.A.2 have been filtered

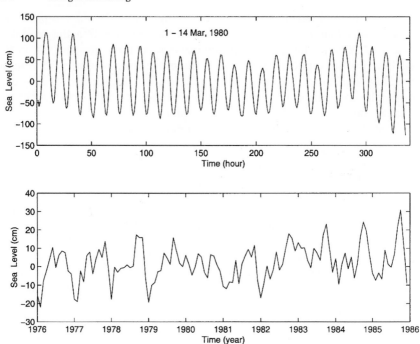

Figure 7.A.1 Examples of sea level at Charleston, in various forms. Upper panel shows the actual hourly values, in which the main feature that emerges is the tidal cycle. Lower panel shows monthly mean values. In this presentation the annual cycle is dominant, but irregular (mostly wind-driven) features are still apparent.

to remove the daily tides out to periods of about 30 h, but to retain power at periods longer than 72 h. The literature on the subject of wind-driven coastal currents is enormous; see, for example, Csanady (1982), Mitchum and Clarke (1986), or Clarke and Van Gorder (1986). We wish to emphasize that the decadal signal of roughly 10 cm/century, while an overwhelmingly significant feature over many years, is relatively small when viewed from a high-frequency perspective. The filtering operations must be done carefully. It is remarkable that coastal tide gauge measurements are sufficiently accurate to allow these small signals to emerge from the background of noise and other effects. Careful analysis of the way the tidal system is kept in calibration, however, shows that this level of accuracy is built into the methodology (Mitchum, 1994).

In examining the lowest frequency signals it is usually adequate to ignore the effects of atmospheric pressure fluctuations because these are small at long periods. At the time scale of a few days to a few weeks, as weather systems pass over a region, the effects of atmospheric pressure signals cause substantial sea level variability, and so must be taken into account. Yet for

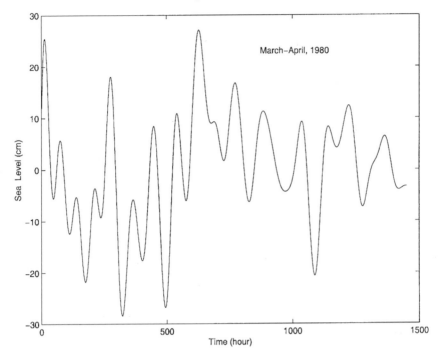

Figure 7.A.2   Sea level at Charleston, as in Fig. 7.A.1. The main feature here is the wind-forced response associated with local oscillatory longshore flows. The signal here shows the energy only at periods longer than 30 hours, suppressing the tidal variations.

periods as long as a year these signals tend to be small at low latitudes and grow larger at higher latitudes. Obviously, a great deal of signal processing takes place before the long-term rise of sea level of Fig. 7.1 becomes evident. It must be done carefully. Small nonlinear terms cannot be safely neglected.

REFERENCES

Barnett, T. P. (1984). The estimation of "global" sea level change: A problem of uniqueness. *J. Geophys. Res.* **89** (C5), 7980–7988.

Chelton, D. G., and Davis, R. E. (1982). Monthly mean sea level variability along the west coast of North America. *J. Phys Oceanogr.* **12,**757–787.

Clarke, J. (1992). Low frequency reflection from a non-meridional eastern ocean boundary and the use of coastal sea level to monitor eastern Pacific equatorial Kelvin waves. *J. Phys Oceanogr.* **22,** 163–183.

Clarke, A. J., and VanGorder, S. (1986). A method for estimating wind-driven frictional time-dependent stratified shelf and slope water flow. *J. Phys Oceanogr.* **16,** 1011–1026.

Csanady, G. T. (1982). *Circulation in the Coastal Ocean.* D. Reidel, Boston.

Cummins, P. F., Mysak, L. A., and Hamilton, K. (1986). Generation of annual Rossby waves in the north Pacific by the wind stress curl. *J. Phys Oceanogr.* **16,** 1179–1189.

Douglas, B. C. (1991). Global sea level rise. *J. Geophys. Res.* **96**, (c4), 6981–6992.

Douglas, B. C. (1992). Global sea level acceleration. *J. Geophys. Res.* **97** (C8), 12,699–12,706.

Ezer, Tal, Mellor, G. L., and Greatbatch, R. J. (1995). On the interpentadal variability of the North Atlantic Ocean: Model simulated changes in transport, meridional heat flux and coastal sea level between 1955–1959 and 1970–1974. *J. Geophys. Res.* **100** (C6), 10,559–10,566.

Greatbatch, R. J., Fanning, A. F., Goulding, A. D., and Levitus, S. (1991). A diagnosis of interpentadal circulation changes in the North Atlantic. *J. Geophys. Res.* **96** (C4), 22,009–22,023.

Hicks, S. D., and Crosby, J. E. (1974). *Trends and Variability of Yearly Mean Sea Level 1893–1972.* Technical Report, U.S. Dept. of Commerce, NOAA, National Ocean Service.

Hicks, S. D., Debaugh, H. A., and Hickman, L. E. (1983). *Sea level variations for the United States 1855–1980,* Technical Report, NOAA, NOS, Silver Spring, MD.

Hong, B. G., Sturges, W., and Clake, J. (2000). Sea level on the U.S. east coast: Decadal variability caused by open ocean wind curl forcing. *J. Phys. Oceanogr.* **29**, in press.

Kessler, W. S. (1990). Observations of long Rossby waves in the northern tropical Pacific. *J. Geophys. Res.* **95** (C4), 5183–5217.

Levitus, Sydney (1990): Interpentadal variability of steric sea level and geopotential thickness of the North Atlantic Ocean, 1970–1974 versus 1955–1959. *J. Geophys. Res.* **95** (C4), 5233–5238.

Mayer, D. A., and Weisberg, R. H. (1993). A description of COADS surface meteorological fields and the implied Sverdrup transports for the Atlantic ocean from 30°S to 60°N. *J. Phys. Oceanogr.* **23**, 2201–2221.

Meyers, G. (1979). On the annual Rossby wave in the tropical North Pacific Ocean. *J. Phys. Oceanogr.* **9**, 663–674.

Mitchum, G. T., and Clarke, J. (1986). Evaluation of frictional, wind forced long wave theory on the west Florida Shelf. *J. Phys Oceanogr.* **16**, 1027–1035.

Mitchum, G., Kilonsky, B., and Miyamoto, B. (1994). Methods for maintaining a stable datum in a sea level monitoring system. *Proc. OCEANS '94 OSATES,* Institute of Electrical Engineering/Ocean Engineering Society, Parc de Penfield, Brest, France.

Platzman, G. W. (1968). The Rossby wave. *Q. J. Roy. Met. Soc.* **94**, 225–246.

Price, J. M., and Magaard, L. (1986). Interannual baroclinic Rossby waves in the midlatitude North Atlantic. *J. Phys. Oceanogr.* **16**, 2061–2070.

Qui, Bo, Miao Weifeng, and Müller, Peter (1997). Propagation and decay of forced and free baroclinic Rossby waves in off-equatorial oceans. *J. Phys. Oceanogr.* **27**, 2405–2417.

Roemmich, D. (1990). Sea level and the thermal variablity of the oceans. In *Sea-Level Change,* pp. 208–220. National Academy Press, Washington, DC.

Slutz, R. J., Woodruff, S. D., Jenne, R. L., and Steurer, P. M. (1985). A comprehensive ocean–atmosphere data set. *Bull. Am. Meteor. Soc.* **68** (10), 1239–1250.

Sturges, W., and Hong, B. G. (1995). Wind forcing of the Atlantic thermocline along 32°N at low frequencies. *J. Phys. Oceanogr.* **25**, 1706–1715.

Sturges, W., Hong, B. G., and Clarke, A. J. (1998). Decadal wind forcing of the North Atlantic subtropical gyre. *J. Phys. Oceanogr.* **28**, 659–668.

White, W. B. (1977). Annual forcing of baroclinic long waves in the tropical North Pacific Ocean. *J. Phys. Oceanogr.* **7**, 50–61.

Woodworth, P. L. (1990). A search for accelerations in records of European mean sea level. *Int. J. Climatology* **10**, 129–143.

Wunsch, C. (1992). Decade-to-century changes in the ocean circulation. *Oceanography* **5** (2), 99–106.

Chapter 8 | Social and Economic Costs of Sea Level Rise

Stephen P. Leatherman

8.1 INTRODUCTION

Coastal zones have been the wellspring of civilization since antiquity because of the ease of water-based transportation. Today about 50% of the world's population lives in this critical interface between land and water; 13 of the world's 20 largest cities are located on the coast. But increasing populations and development are placing significant stresses on coastal resources; rising sea level is causing land loss, creating a collision course of social and sea level trends (Fig. 8.1).

Changing sea levels have been the driver of shorelines over geologic time. For example, tens of millions of years ago, the seas had advanced several hundred kilometers farther inland than today along the U.S. east coast. These ancient shorelines are marked by the "falls" in the river systems, and many major cities and state capitals are located there (e.g., Washington, DC; Baltimore, Maryland; Richmond, Virginia; and Raleigh, North Carolina). The fall line is the furthest upstream position for navigable waters, and delimits the inland edge of the coastal plain. Consider that the Mall in Washington, DC, is little more than 3 m above sea level, and the Potomac River there is tidally driven. It is interesting that all the water tied up in the world's glaciers and ice caps has a sea level equivalent of about 70 m. Obviously if even a fraction of this ice melts or slumps into the sea, the consequences will be severe for low-lying coastal cities.

Sea levels have been much lower in the recent geologic past, approaching 125 m below present levels approximately 21,000 years ago. Coastal resorts, like Atlantic City, New Jersey, and Virginia Beach, Virginia, would have been several hundred kilometers landward of the shoreline at that time as the seas receded to the shelf edge, exposing the wide continental shelf. The Chesapeake Bay and associated salt marshes did not exist then; the Susquehanna River was a rapidly flowing river that reached all the way to the shelf edge.

Unlike most other manifestations of climate change, sea level rise is already a problem, and the impacts on coastal regions are easily seen (Figs. 8.2 and 8.3). Global sea level has risen nearly 20 cm in the last century (Douglas,

Sea Level Rise

181

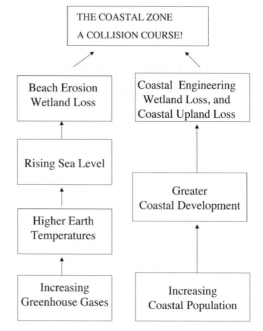

Figure 8.1   The coastal zone is increasingly becoming a collision course.

Figure 8.2   Episodic storms cause dramatic damage, but slowly rising sea levels cause gradual beach erosion that puts houses and infrastructure at risk. Coast Guard Beach, Cape Cod, Massachusetts.

Figure 8.3   Fire Island along the South Shore of Long Island, New York, after the December 10–12, 1992 northeaster.

1991, 1997; Peltier and Jiang, 1997; Peltier, Chapter 4 of this volume), and there is clear physical evidence of impacts manifested by worldwide coastal land loss. For instance, over 70% of the world's sandy beaches are currently eroding (Bird, 1985), and the percentage increases to 80 to 90% for the better-studied and better-documented U.S. sandy coasts (Leatherman, 1986; Galgano, 1998). While many factors contribute to shoreline recession, sea level rise is considered the underlying factor accounting for the ubiquitous coastal retreat (Vellinga and Leatherman, 1989; Zhang, 1998; this chapter).

Accelerated sea level rise is usually regarded as the most certain consequence of global warming (IPCC, 1996), with many adverse consequences for the world's coastal zone. Rising sea levels erode beaches by increasing offshore and longshore loss of sediment, directly inundating (or submerging) marshes and other low-lying lands, increasing the salinity of estuaries and aquifers, raising coastal water tables, and exacerbating coastal flooding and storm damage (Fig. 8.4). Beach erosion along developed coasts is the most obvious problem of rising sea level, but wholesale losses of coastal wetlands can also be expected. The coastal zones most vulnerable in terms of habitable land area are small islands and river deltas. There is the potential for displacement of millions of people.

Assessments of economic impacts for industrialized nations indicate that the cost for a 1-m rise is high, but manageable. For instance, best estimates for coastal stabilization are $12 billion for the Netherlands, $74 billion for

- Erosion of beaches and bluffs
- Inundation of low-lying areas
- Salt intrusion into aquifers and surface waters
- Higher water tables
- Increased flooding and storm damage

Figure 8.4   Physical effects of sea level rise.

Japan (IPCC, 1992), and as much as $200 to $475 billion for the United States (Titus *et al.,* 1991). Much less work has been undertaken in developing nations; Nicholls *et al.* (1995) have compiled the available data (Table 8.1). Many of the most vulnerable nations have large and increasing populations and yet often lack the resources and institutional framework to deal with contemporary impacts and problems.

The world's coastline is about a million kilometers in length and consists of a wide variety of landforms, from cliffs and headlands to barrier beaches and coastal plains. Rising sea levels will potentially impact all coastal environments to varying degrees, but sandy beaches, coastal wetlands, deltaic coasts, and small islands will be the most affected. Beaches are the number one tourist destination worldwide, and erosion is already threatening much of the expensive beachfront development. The ecological and economic value of coastal wetlands is now appreciated, but human impacts have taken a heavy toll, and sea level rise portends even more severe consequences in the future. Deltas, for example, exist in delicate balance with sea level, and human settlements located therein are vulnerable to even small increases of sea level. Finally, low-lying coral reef islands are the most at risk since entire island nations could vanish during the next century.

8.2 DATA CONSIDERATIONS

The most vexing problem for determining the impact of a forecast rise in 21st-century sea level of 0.5 m or more (IPCC, 1996), is the lack of accurate data on elevations for most of the world's coastal zones. The most commonly available topographic maps for the U.S. coasts only delimit the 3-m contour line, which may be located a substantial distance (hundreds to thousands of meters) landward of the shoreline on low-lying coastal plains. Elsewhere, many maps in developing countries have contour intervals of only 10 to 50 m, making them virtually worthless for use in analyzing impacts of any reasonable sea level rise scenario.

Table 8.1

Socioeconomic Impacts of Sea Level Rise (Nicholls *et al.*, 1995)

Country	People affected		Capital value at loss		Land at loss		Wetland at loss (km²)	Adaptation costs/protection	
	100s	% total	US $ (million)	% GNP	km²	% total		US $ (million)	% GNP
Antigua	38	50	–	–	5	1.0	3	76	0.32
Argentina	–	–	5600[3]	6	3400	0.1	1100	>1800	>0.02
Bangladesh	71,000	60	–	–	25,000	17.5	5800	>1000[5]	>0.06
Belize	70	35	–	–	1900	8.4	–	–	–
Benin	1350	25	126	12	230	0.2	85	>430	>0.41
China	72,000	7	–	–	35,000	–	–	–	–
Egypt	4700	9	59,272	204	5800	1.0	–	13,133[6]	0.45
Guyana	600	80	4000	1115	2400	1.1	500	200	0.26
India	7100[2]	1	–	–	5800	0.4	–	–	–
Japan	15,400	15	807,000	72	2300	0.6	–	>159,000	>0.12
Kiribati[1]	9	100	2	8	4	12.5	–	3	0.10
Malaysia	–	–	–	–	7000	2.1	6000	–	–
Marshall[1] Islands	20	100	175	324	9	80	–	>380	>7.04
Mauritius	6	1	–	–	10	0.5	–	–	–
The Netherlands	10,000	67	186,000	69	2165	5.9	642	12,286	0.05
Nigeria	3200[2]	4	18,000[3]	52	18,600	2.0	16,000	>1400	>0.04
Poland	235	1	24,000	24	1700	0.5	36	1500	0.02
Senegal	110[2]	>1	700	14	6100	3.1	6000	>1000	>0.21
St. Kitts-Nevis[1]	–	–	–	–	1	1.4	1	53	2.65
Tonga[1]	30	47	–	–	7	2.9	–	–	–
Uruguay	13[2]	<1	1800[3]	26	96	0.1	23	>1000	>0.12
United States	–	–	–	–	316,000[4]	0.3	17,000	>143,000	>0.03
Venezuela	56[2]	<1	350	1	5700	0.6	5600	>1700	>0.03

[1] Minimum estimates—incomplete national coverage.
[2] Minimum estimates—number reflects estimated people displaced.
[3] Minimum estimates—capital value at loss does not include ports.
[4] Best estimate is that 20,000 km² of dry land is lost, but about 5,400 km² are converted to coastal wetlands.
[5] Adaptation for Bangladesh only provides protection against a 1-in-20 year event.
[6] Adaptation costs for Egypt include development scenarios.

There are a number of methods for obtaining topographic information, but the quality obtained and cost involved vary widely. Traditional photogrammetric approaches are generally too time consuming and expensive for regional or national analysis and have application only to localized areas of interest. Available satellite imagery is only accurate to within about 10 m vertically under the best conditions (Theodossiou and Dowman, 1990).

Leatherman and Nicholls (1995) developed a reconnaissance-level assessment of probable land loss in response to sea level rise called AVVA (aerial videotape-assisted vulnerability analysis). This approach involves a combination of unrectified oblique aerial video recording of the coastline from very low altitude (e.g., 50–100 m), limited ground truth information, archival research, and analysis of these data in conjunction with simple land loss and response models. Videotaping the coastline from a small plane at low altitude captures the relative aspect of the land to the sea, coastal geomorphology, and land use. From the videotapes and limited ground surveying, it is possible to estimate approximate coastal elevations near the shoreline. Clearly, there can be large errors associated with this technique, but controlled studies in the Chesapeake Bay, Maryland, showed that such estimates were unbiased (Leatherman and Nicholls, 1995). An aggregated analysis can yield useful results for large-scale studies, but the data are not appropriate for site-specific information because AVVA does not utilize photogrammetric procedures.

The aerial video record is especially useful for determining coastal geomorphology and land use in developing countries. Many of the national studies conducted in developing countries have utilized the AVVA procedure to conduct land loss analyses (IPCC, 1992). Table 8.2 lists the different elements of such a classification system. Additional information should be collected to supplement the video record, including historical photographs and other data (Table 8.3).

Topographic survey technology has greatly advanced in the last few years with the advent of the airborne laser terrain mapper (Carter *et al.,* 1998). The airborne laser system (also called LIDAR) can be operated from a light single- or twin-engine airplane. Data are taken while flying at 150 to 200 km/h at altitudes between 500 and 1000 m with a scan angle of 10 to 20° and a laser repetition rate of 5000 to 25,000 pulses/s. With this system, it is possible to cover an area several hundred meters in width and hundreds of kilometers in length in a few hours. The laser point (or "footprint") spacing can be a few decimeters to a few meters. Differential GPS navigation of the aircraft enables height measurements in a geodetic reference system accurate from 5 to 10 cm, and detection of topographic features with height differences between 2 and 5 cm. This incredible new technology, now operational, will literally revolutionize our surveying capability and make possible definitive impact analyses of any reasonable sea level rise scenario. A digital videocamera can also be aimed through the same borehole as the laser, providing much of the same type of information obtained in the AVVA reconnaissance technique.

Table 8.2
Geomorphology and Land Use Classification System

I. Coastal geomorphology
  A. Beaches
    1. Barrier beach
      a. Type
        (i) Bay barrier
        (ii) Barrier spit
        (iii) Barrier island (microtidal or mesotidal)
      b. Morphology
        (i) High (>5 m), continuous foredune
        (ii) Extensive dune field
        (iii) Low dunes with washovers
    2. Strandplain or headland beach
      a. Type
        (i) Low coastal plain
        (ii) Flanked by erodible cliffs
        (iii) Flanked by hard rock cliffs
      b. Morphology
        (i) High (>5 m), continuous foredune
        (ii) Extensive dune field
        (iii) Low dunes with washovers
    3. Pocket beach
      a. Type
        (i) Flanked by erodible headlands
        (ii) Flanked by rocky headlands
      b. Morphology
        (i) High (>5 m), continuous foredune
        (ii) Extensive dune field
        (iii) Low dunes with washovers
  B. Wetlands
    1. Estuary
      a. Mangrove
      b. Marsh (grass)
      c. Marsh (scrub shrub)
      d. Marsh (forested)
    2. Delta
      a. Mangrove
      b. Marsh (grass)
      c. Marsh (scrub shrub)
      d. Marsh (forested)
    3. Backbarrier areas
      a. Mangrove
      b. Marsh (grass)
      c. Marsh (scrub shrub)
      d. Marsh (forested)
    4. Tidal flats
  C. Cliffs (no beach)
    1. Erodible
      a. Dunes above cliff
      b. Flatland above cliff
      c. Hilly land above cliff
      d. Mountainous land above cliff
      e. Height
    2. Rocky
      a. Dunes above cliff
      b. Flatland above cliff
      c. Hilly land above cliff
      d. Mountainous land above cliff
  D. Muddy coast
      a. Flatland behind mud beach
      b. Hilly land behind mud beach
      c. Mountainous land behind mud beach
      d. Lake behind mud beach
      e. Lagoon behind mud beach
  E. Hardened (protected) shoreline
      a. Sand dunes behind protection
      b. Flatland behind protection
      c. Hilly land behind protection
      d. Mountainous land behind protection
      e. Wetland behind protection
II. Protection (if present)
  A. Seawall
  B. Bulkhead (composition: timber, concrete, riprap, etc.)
  C. Breakwater
  D. Groins
  E. Jetty
  F. Protected Harbor
  G. Beach Nourishment
III. Land use
  A. Urban/City.
  B. Residential
  C. Industrial
  D. Tourist
  E. Agricultural (crop types)
  F. Cattle grazing
  G. Sheep grazing
  H. Orchards
  I. Forest
  J. Barren
  K. Shrub lands
  L. Desert
  M. Fishing
  N. Aquaculture
IV. Inland geomorphology
  A. Flatland
  B. Hilly land
  C. Mountainous land
  D. Lake
  E. Wetlands

Table 8.3

Collection of Existing Coastal Data

The existing data are often scattered, and compiling all the information described below can require a considerable effort. Further some of these data will not be available. However, once assembled, a coastal data base provides a basis for more informed decisions and helps to identify critical data gaps.

1. Maps showing what activities are located on the coast, within 6-m (20-foot), 3-m (10-foot), and 1-m-(3 foot) elevation of mean sea level such as (a) cities; (b) ports; (c) resort areas and tourist beaches; (d) industrial areas; and (e) agricultural areas.
2. Best topographic maps, ideally with at most a 3-m (10-foot) contour interval.
3. Survey maps.
4. Aerial photographs.
5. Satellite images (as available).
6. Tide tables.
7. Long-term relative sea level rise data (obtained from the Permanent Service for Mean Sea Level at the Proudman Oceanographic Laboratory, England,* as well as local sources).
8. Other historical data.
   (a) The magnitude and damage caused by flooding from maps, photographs (aerial or other), and other sources.
   (b) Accurate coastal maps from different years to examine coastal changes, such as beach erosion.
   (c) Photographs from different years to show coastal changes such as beach erosion or variability.
9. Population density and other demographic data

Four impacts of sea level rise are considered below. Their relative importance will vary from site to site, depending on a range of factors.

1. Increased storm flooding
   (a) Describe what is located in flood-prone areas.
   (b) Describe historical floods, including
       (i) location, magnitude and damage,
       (ii) response of the local people, and
       (iii) response of government. How have policies towards flooding evolved?
2. Beach/bluff erosion.
   (a) Describe what is located within 300 m (1,000 feet) of the ocean coast.
   (b) Describe beach types.
   (c) Describe the various livelihoods of the people living in coastal areas such as commercial fishers or international based coastal tourism. Note subsistence lifestyles such as coastal fishermen.
   (d) Describe any existing problems of beach erosion including quantitative data. These areas will experience more rapid erosion given accelerated sea level rise.
   (e) For important beach areas, conduct a Bruun rule analysis and preferably a trend analysis to assess likely shoreline recession. What existing coastal infrastructure might be impacted by such recession?
3. Wetland and mangrove inundation and loss.
   (a) Give a comprehensive description of wetland areas, including human activities and resources which depend on the wetlands. For instance, are mangroves being cut and used? Do fisheries depend on wetlands?
   (b) Have wetlands or mangroves been reclaimed for other uses, and is this likely to continue?

* http://www.pol.ac.uk./psmsl/psmsl.info.html.

8.3 SANDY BEACH EROSION

Sea level is one of the principal determinants of shoreline position. There is a causal relationship among several factors: sea level, sediment supply, wave energy, and shoreline position. Rising sea level should increase beach erosion, except where this trend is offset by abundant sediment supply.

Sea level rise induces beach erosion or accelerates ongoing shore retreat for several reasons (Leatherman, 1991). First, higher water level enables waves to break closer to shore. Second, deeper water decreases wave refraction and thus increases the capacity for longshore transport. Finally, with higher water level, wave and current erosion processes act farther up the beach profile, causing a readjustment of that profile. Maintenance of an equilibrium beach/ nearshore profile in response to sea level rise requires an upward and landward displacement of the beach in time and space; this translates to erosion in ordinary terms.

There is considerable confusion in the literature regarding the difference between erosion and inundation, particularly since both cause land loss. Often these terms are used interchangeably, but there are important differences. Erosion is the physical removal of material by waves and currents from the beach profile. When the beach erodes, sediment is either lost offshore beyond closure depth or lost longshore to sinks such as a inlets and lagoons/bays. Inundation in contrast is merely the permanent submergence of low-lying land and does not imply any sediment movement per se. Such low-lying coastal lands along sheltered coasts experience periodic flooding during storm events, and even spring high tides, prior to permanent inundation. Total coastal retreat or recession equals the sum of erosion plus inundation. Along open-ocean beaches, over 90% of the retreat is caused by erosion; the opposite is generally true for coastal marshes in sheltered bays, lagoons, and estuaries with limited wave action.

Most sandy shorelines worldwide have retreated during the past century (Bird, 1976; NRC, 1990). Progradation or coastal accretion is restricted to areas where locally abundant sediment is supplied by rivers or where the land is being elevated by isostatic glacial rebound or tectonic uplift. The ubiquitous recession of shorelines cannot be attributed to an increase in storminess because such a storm trend would have to be worldwide in extent. Human interference in the form of engineering changes to the shore can locally account for exceedingly high erosion rates, but cannot be considered a primary cause because erosion is occurring also on sparsely populated and little-developed sandy coasts (Leatherman, 1991). One phenomenon that is both very widespread and physically associated with erosion is global sea level rise.

Bruun (1962) was the first person to posit a role for rising sea level in shore erosion. His argument is largely based on the concept of an equilibrium profile, which refers to a statistical average profile at a particular location that maintains its form apart from fluctuations. Thus the equilibrium profile is a conceptual model that represents shoreline behavior averaged in time over both seasonal changes, and the longer period, storm-induced erosion/recovery cycles that can take years to complete. However, attempts to correlate coastal erosion with sea level rise have not been convincing until recently because of the difficulty of obtaining accurate long-term shoreline position data, and problems from other factors, such as human impacts. Most analyses have pointed to a preponderance of local reasons for beach erosion, rather than a common explanation; any sea level rise contributions have been effectively masked by other, more significant, mechanisms on a site-specific basis (IPCC, 1996).

The Bruun model is a deceptively simple two-dimensional balancing of sediment in the on- and offshore direction in response to sea level rise. The beach profile achieves a new equilibrium by shifting landward and upward, resulting in erosion of the beach and nearshore area and deposition on the lower part of the beach profile (Fig. 8.5). What the model predicts is that long-term sandy beach erosion will be several orders of magnitude larger than the rate of sea level rise. This relation of sea level rise to erosion has become known as the *Bruun rule,* and it is widely cited. It is very important to note that the Bruun model does not suggest that sea level rise actually causes erosion; rather, increased sea level enables high-energy, short-period storm waves to attack farther up the beach and transport sand offshore. While this model has been shown to apply in wave tanks and some lake environments (Hands, 1983), studies along the open-ocean coast have been limited, and past results incomplete and controversial. Some work seems to verify it, while others allege that the Bruun model does not work at all (SCOR, 1991).

The controversy surrounding the Bruun erosion model comes from its simplicity compared to actual beach processes. Sandy beaches are highly dynamic features, making it difficult to observe a small underlying trend of erosion. Tremendous fluctuations in beach width (and hence shoreline posi-

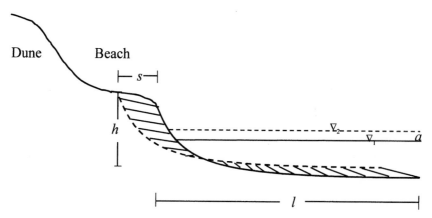

Figure 8.5 Beach response to sea level rise based on the Bruun rule (not to scale). Note that the model does not suggest that sea level rise actually causes the erosion; rather, increased sea level enables high-energy, short-period waves to attack farther up the beach and transport sand offshore.

tion) are caused by changes in sediment supply, seasonal wave energy variation, storm impact, and human interference on a temporal and spatial basis. Historical shoreline data spanning more than 80 years and preferably longer are needed to obtain even approximate long-term trend estimates of beach erosion (Galgano *et al.,* 1998). This is so for three reasons. First historical shoreline position data are inherently accurate to only about 8–9 m (Crowell *et al.,* 1991), and even modern methods will have a noise level of at least a few meters because of the imprecision of the definition of a shoreline indicator. Second, a great storm can cause more erosion in a few hours or days than may have occurred in the previous 50 years. Finally, it can take beaches more than a decade to recover from major storm impact. Examples of such an extended recovery time are the Ash Wednesday Storm of 1962 along the Delaware Atlantic coast (Galgano *et al.,* 1998) and the landfall of Hurricane Alicia in 1983 at Galveston Island, Texas (Morton *et al.,* 1994). Thus comparisons of erosion trends with sea level rise to ascertain the usefulness of the Bruun model require erosion rates computed from very long shoreline position records. The necessary shoreline position data covering near-century or longer spans are obtainable in the United States from historical and modern National Ocean Survey (NOS) T-sheets, vertical aerial photography, and kinematic GPS surveys (now from LIDAR as well). These data have the requisite high accuracy (Crowell *et al.,* 1991), length of record (100+ years), and spatial characteristics (entire U.S. mid-Atlantic coast) to enable computation of the underlying erosion trend.

The U.S. east coast is an ideal location to test the predictions of the Bruun model. Sea level rise varies by about a factor of 2 along the coast because of the effects of glacial isostatic adjustment (see Chapter 4), so there should be a corresponding geographic variation of long-term erosion rates. This coast is also very well covered by tide gauge measurements, so that reliable estimates of the 20th-century rate of sea level rise are available. Douglas (1991) showed that at least 50 years of record are required to obtain a stable, long-term rate estimate. Therefore, in the discussion that follows, only tide gauges with record lengths of 50 years or more along the U.S. east coast were selected to determine the long-term rate of sea level rise. It is interesting that geographic variations of wave energy have been shown to have no effect on long-term shoreline change (Zhang, 1998).

To obtain erosion rates for comparison to trends of sea level, I and my colleagues K. Zhang and B. Douglas (2000) computed long-term (100+ year) shoreline change rates by linear regression at 100-m intervals along the U.S. east coast for five coastal compartments. These are Long Island (New York), New Jersey, Delmarva (i.e., Delaware, Maryland, and Virginia), North Carolina, and South Carolina. Three methods were used to compute these change rates: (1) averaging of all shoreline rates for each compartment regardless of any morphological details or differences, (2) averaging of shoreline change rates over both erosional and accretional sections not influenced by inlets and coastal engineering projects, and (3) averaging shoreline change rates for erosional sections alone not influenced by inlets and coastal engineering projects. The results of these analyses are presented in Table 8.4.

Beaches affected by inlets and coastal engineering projects display large spatial variations in long-term shoreline change rates (Table 8.4, column 2). It is well known that inlets have a dominant influence on the majority of U.S. east coast beaches (Galgano, 1998). Accretion occurs updrift of inlets, with severe erosion downdrift. Some shorelines away from inlets exhibit natural stability or slight accretion because of gradients in longshore sediment transport, or local onshore sediment transport. Erosion "hot spots" also exist. All of these anomalous areas were eliminated from the data set used to evaluate the Bruun model because it only applies to beaches with no net change of sediment supply (see Table 8.4, column 5). For this selected data set, there is remarkably good correlation between sea level rise and average long-term shoreline change for eroding beaches (Fig. 8.6). The high correlation ($r^2 = 0.89$) indicates a very strong association between sea level rise and beach erosion. However, the ratios of shoreline change rate versus sea level rise rate vary from 110 to 181 (Table 8.4, column 5), larger on average than 50 to 200 according to Bruun's original calculation. The significant variability observed probably represents local geological differences. But what is confirmed is that the lateral beach erosion rate is always *two orders of magnitude or more than the rate of sea level rise!* Forcing the regression to pass through the origin (zero erosion for zero sea level rise) yields an average shoreline change rate response that is about 150 times the sea level rise rate (Fig. 8.6). This analysis shows that future sea level rise will cause retreating beaches to continue to erode, accelerated sea level rise will increase the rate of erosion along such coasts, and stable to slightly accretional shorelines should begin to erode in the future unless additional, excessive sediment supplies are locally available.

Table 8.4

Long-Term Shoreline Change Rates (m/yr) along the U.S. East Coast

Compartment	Entire compartment	Areas not influenced by inlets and coastal engineering projects	s/a*	Erosional areas not influenced by inlets and coastal engineering projects	s/a
Long Island	−0.44 ± 0.89	−0.13 ± 0.25	53	−0.27 ± 0.17	110
New Jersey	−0.12 ± 1.5	−0.38 ± 0.58	101	−0.68 ± 0.36	181
Delmarva	−2.26 ± 3.39	−0.20 ± 0.74	56	−0.53 ± 0.35	150
North Carolina	−0.87 ± 1.79	−0.32 ± 0.41	88	−0.50 ± 0.25	139
South Carolina	−0.78 ± 2.73	−0.34 ± 0.38	109	−0.41 ± 0.35	132

* *s/a* represents the ratio of shoreline retreat rate to sea level rise rate.

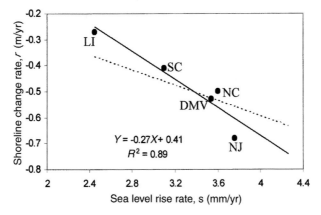

Figure 8.6   Relationship between rates of sea level rise and shoreline retreat for erosional areas not influenced by inlets or coastal engineering projects. The solid line is the best fit; the dashed line passes through the origin.

Nearly all past studies projecting shoreline positions and erosion rates in response to sea level rise have been based on the Bruun model (Nicholls and Leatherman, 1995a), so it is reassuring that this approach has been validated in an average sense. However, the model assumes that there is a seaward limit of the active beach profile, and also that the entire profile is sandy. Where in fact the seaward extent of the shoreface/inner zone is very large, there is a distance offshore at which the depth is great enough that the bottom is relatively undisturbed, and the Bruun model is a reasonable first approximation of the actual physical situation. However, the depth at which significant sediment motion is absent (often called the *depth of closure*) is a rather vague concept in an oceanic wave environment. The problem is confounded when an attempt is made to quantify a relatively confined zone of erosion (e.g., the narrow beach/dune zone) with a broad zone (the shoreface/ inner shelf) over which eroding sediment can be thinly spread (Leatherman, 1991).

The depth of closure can be estimated by a range of techniques, including grain size trends, orientation of offshore contours, and wave statistics (Hallermeier, 1981). Where wave data are available, this is the best approach since the depth of closure can be related to the time scale of interest. Depth of closure is time dependent; the longer the time period being considered, the larger the depth of closure (Hands, 1983; Stive *et al.*, 1992). Nicholls *et al.* (1995) considered two estimates of depth of closure, which would encompass the actual value. These are the annual depth of closure and the depth of closure over the time frame of a century. The principal limitation is the availability of good-quality wave data. In practice, active profile width must be estimated for most developing countries using expert judgment rather than hard data, which can lead to errors in rates of shoreline recession and therefore the cost of response (e.g., beach nourishment).

Another approach to predicting future shoreline position involves historical trend analysis. This method consists of an empirical determination of past trends and projection of new shorelines using historical trends. Application of this methodology is largely limited to industrialized nations with well-studied beaches and long-term, quantitative shoreline position data. In this case, shoreline response is based on the historical trend with respect to local sea level change during that time period. This procedure in a general way accounts for the inherent variability in shoreline response based on differing coastal processes, sedimentary environments, and coastline exposures (Leatherman, 1991). This straight-line projection of future shoreline position, which is based on the ratio of the rates of beach erosion and sea level rise, assumes that sea level rise is responsible for all the change as opposed to any other cause. The underlying assumption of this analysis is that shorelines will respond in the future as in the past because sea level rise is the driving function and all other parameters remain essentially constant. As noted above, one of the great difficulties in confirming the Bruun model is the pervasive influence of tidal inlets and coastal engineering projects on beach behavior.

There are three general responses to shore erosion: (1) retreat from the shore (Fig. 8.7), (2) armor the coast (Fig. 8.8), or (3) nourish the beach (Fig.

Figure 8.7   Relocation of houses along eroding shores is the most economical approach along nonurbanized beaches, but there is a loss of real estate.

Figure 8.8   The Galveston seawall has withstood the test of time, but at the expense of the beach.

8.9). The choice of a response strategy will depend upon a number of factors, including socioeconomic and environmental conditions. The retreat option is the preferred option for undeveloped or sparsely developed areas. For highly urbanized areas, such as Miami Beach, Florida, or Atlantic City, New Jersey, the abandonment option is not politically realistic or economically viable. The value of this beachfront property is several hundred million dollars per kilometer. For instance the "Gold Coast" of Florida, which includes Palm Beach to Miami Beach, has an appraised value of over $1 trillion. Beach nourishment is necessary to maintain beaches for tourism, but seawalls are also often emplaced to provide a measure of extra protection for the expensive tourist resorts and infrastructure. If the erosion rate is too high or the cost of beach nourishment prohibitive, then progressive erosion will eventually eliminate the sandy beach, resulting in a hardened shoreline. The Galveston, Texas, seawall is the most famous case in point. This almost century-old fortification has stood the test against hurricane landfall and protected a low-lying, vulnerable barrier island development from repeated storm attack (Figure 8.8).

Figure 8.9 Beach nourishment through the onshore pumping of coarse sediment is the preferred approach for communities to maintain their recreational corridor and provide storm protection, but the cost is often high and future renourishment always a necessity.

8.4 COASTAL WETLANDS

Sheltered coasts, such as bays, lagoons, and esturies, often provide the low wave environment necessary for the development of coastal wetlands. In temperate climates, salt marshes dominate the intertidal zone (between low and high tide) and can range up to a meter above the mean high tide level. Most of the coastal wetlands in the United States are dominated by salt marshes (Spartina vegetation) except for South Florida where mangrove stands are found.

Along sheltered coasts, rising sea level often causes direct inundation or submergence of the upland. This impact results in higher water tables and boggy soils, as well as invasion of salt-tolerant species (halophytes) into and displacing farmer's crops. There is also an eventual die-off of trees because of saltwater intrusion. While there can be some edge erosion, land loss is largely a simple function of slope—the smaller the slope, the greater the inundation (Fig. 8.10). High water is often selected as the primary inundation contour, and all land loss is measured relative to this datum. The choice of the high-water line reflects a practical distinction between a lower zone of inundation (daily submergence by the tides) and an upper zone of increased coastal flooding (Nicholls *et al.*, 1995). A higher elevation boundary (e.g.,

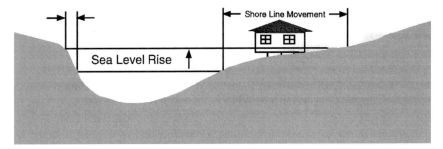

Figure 8.10   Land loss due to inundation is slope dependent.

land inundated by spring or annual storm tides) is more appropriate for distinguishing between usable land for development. For insurance purposes, the U.S. Federal Emergency Management Agency (FEMA) requires that houses be elevated above the 100-year flood level. This can be accomplished by building on stilts or high pilings and enables development close to the water's edge (a location coveted by most people) even though the land itself can be considerably below this storm surge elevation.

Loss of coastal wetlands is already occurring in most countries both for natural reasons and because of human interference, including present sea level rise, reduction in sediment input, saline intrusion, dredging and filling for development, cutting of mangroves for wood fuel, and land reclamation, particularly in Asia. For example, most coastal wetlands in China have already been eliminated through diking and draining in order to provide fertile land for agriculture, and more recently for aquaculture, in order to feed the ever-increasing population. In Bangladesh and Brazil, mangroves have been felled as a primary source of firewood for cooking by the indigenous population of poor people. In peninsular Malaysia, mangrove reclamation for rice production is a governmental imperative, which could mean the loss of nearly all the remaining mangrove ecosystem. In the Niger delta area, upstream dam construction is causing sediment starvation and exacerbating existing problems of wetland loss because of saltwater intrusion (Nicholls *et al.*, 1995). Salt marshes that have historically migrated landward in response to slowly rising sea level are now constrained by embankments in England (IPCC, 1996). This problem is common along the backbarrier bays and lagoons of the U.S. east and Gulf coasts because bulkheads are often used to stabilize the shoreline and provide building sites through dredge and fill operations for waterfront homes. Therefore, any acceleration in sea level rise will severely threaten coastal wetlands throughout the world because of the significant stresses already placed on these biological communities through human interference and manipulation.

Coastal marshes can accrete vertically due to biomass production and/or sediment input and keep pace with slow rates of sea level rise (Gehrels and Leatherman, 1989; Kearney and Stevenson, 1991). Inundation and land loss begin to occur when the rate of sea level rise exceeds some threshold value, which is site-specific and often poorly known. Wetland losses can occur over wide areas within a few decades through the "drowning mechanism" wherein individual plants become waterlogged and sulfates built-up in the soil layer beyond the plant's tolerance level (Stevenson *et al.*, 1986).

Most studies of the impact of sea level rise on tropical mangrove ecosystems have been primarily of an historical nature (Woodroffe, 1990). Extensive mangrove stands became established in Australia and elsewhere about 6000 years ago, when the rate of sea level rise slowed down appreciably (IPCC, 1996). Ellison and Stoddart (1991) stated that mangroves located in areas with low sediment input are unable to accrete vertically fast enough to compensate for projected sea level rise, resulting in ecosystem collapse for some island environments, such as Bermuda. Evidence from the Florida Keys (Snedaker *et al.*, 1994), however, indicated that low-island mangroves may be much more resilient than originally thought and may be able to withstand twice the upper limits suggested by Ellison and Stoddart (1991).

Given the complexity of coastal wetlands and the need for reconnaissance-level analyses to estimate the global implications of sea level rise, a simple modeling approach is required for predictive purposes (Fig. 8.11). This approach is based on two threshold rates of sea level rise—a lower threshold rate below which no wetland loss occurs, and an upper threshold which delimits total loss. Between these threshold values, losses are linearly modeled in the

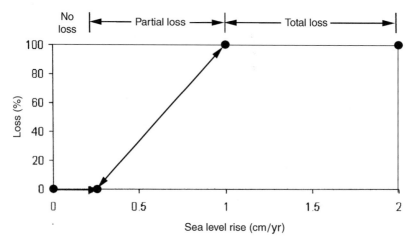

Figure 8.11   Wetland response to sea level rise, assuming a lower threshold value of 0.2 cm/yr and an upper threshold of 1.0 cm/yr (from Nicholls *et al.*, 1995).

absence of any better site-specific information. The initial threshold value is taken as the present rate of relative sea level rise, and any acceleration in this rate is predicted to cause some wetland loss. Ecosystem breakdown was assumed for the 1-m rise scenario (or about 1 cm/yr). These assumptions are not valid in areas of high terrigenous sediment input (e.g., some riverine and deltaic environments) or where organic production of sediment is high relative to the local rate of sea level rise. There are some coastal wetlands in South America where there is no reported land loss in undisturbed, rural areas (IPCC, 1996). In any case total loss is unlikely due to the conversion of low-lying upland areas to coastal wetlands through landward migration as sea level rises, except where bounded by naturally high scarps or embankments and bulkheads as noted above.

Coastal wetlands account for much of the land less than 1 m above sea level. These extensive marshes, swamps, and mangrove forests fringe most of the U.S. coastline, particularly along the Atlantic and Gulf coastal plains. Coastal wetlands, while hard to evaluate economically, serve as nurseries for fish and shellfish, many birds, and fur-bearing animals. They are vital to coastal recreation, to the maintenance of water quality, and as a buffer against shore erosion.

It is estimated that originally there were over 5 million acres of salt marshes in the United States. More than half of these vital wetland areas have been lost, largely due to dredge and fill operations for urbanization. For example, the main part of Boston University, north of Commonwealth Avenue, is built on tidal flats, and land filling dates back to 1794. Many other coastal cities are built on land that was formerly marshy, including the downtown area of the world famous Waikiki Beach in Honolulu, Hawaii.

8.5 DELTAIC COASTS

Deltas form where terrigenous sediment brought down to the coast by rivers accumulates more rapidly than removed by waves and currents. There is a wide spectrum of deltaic coasts around the world (Wright, 1978), but all are the result of interaction between fluvial and marine processes in different oceanic settings. Therefore, deltaic coasts are particularly vulnerable to any diminishment of sediment supply because of dams or dikes and/or any acceleration in the rate of sea level rise.

Deltas are considered to be more vulnerable to sea level rise than any other coastal areas because they comprise extensive low-lying land that is often heavily populated. Since ancient times, deltas have served as the bread-basket for countries like Egypt because of the fertile soil and plentiful fresh water for agriculture. Other deltaic areas became important areas for the survival of civilizations; the Germanic invasion during Roman times pushed the indigenous peoples onto the mudflats and marshes of the Po delta to establish what was to become the grand city of Venice, Italy. Fisheries are also abundant in the fresh and brackish waters of deltaic environments, and the main river channels serve as good navigable waterways, further promoting the development of vast cities in many countries on deltaic coasts.

Broadly based studies have been undertaken in developing countries to identify the vulnerability of deltaic areas to sea level rise, particularly the heavily populated Nile delta in Egypt and the Ganges-Brahmaputra delta in Bangladesh (Milliman *et al.,* 1989). These areas have enhanced vulnerability to sea level rise due to (1) high rates of coastal land subsidence, both naturally occurring and artificially induced and (2) actual or potential sediment starvation caused by upstream dams such as the Aswan High Dam on the Nile River. Human activities, such as groundwater and petroleum extraction, dam building, and construction of flood prevention structures (e.g., dikes) within the delta, exacerbate this natural vulnerability. Therefore, deltaic coasts are already vulnerable to present relative sea level rise even without climate change.

A series of case studies of deltaic areas based on the 1-m rise scenario was undertaken by Nicholls and Leatherman (1995a). In the Nile delta of Egypt, 12 to 15% of the existing agricultural land is at risk from inundation, and over 6 million people at current population levels potentially could be displaced, including half the population of Alexandria. Urbanization of this near–sea level land is still occurring at a rapid pace, which will put considerably larger numbers of people at risk in the coming decades as the water level continues to rise. In Bangladesh, 16% of the land currently used for rice production could be inundated, displacing 13 million people at the same time (Fig. 8.12). In China, 72 million people and tens of thousands of square kilometers of prime agricultural land are at risk. The Chinese, however, are committed to protecting their populated areas from flooding by building and raising existing dikes, such as those that currently protect Shanghai, located on one of the four major coastal deltaic plains.

The potential for displacement of millions of people from their homes by sea level rise, creating a major environmental refugee problem, was a principal factor in raising consciousness about the potentially adverse consequences of global change. Nicholls and Leatherman (1995b) avoided use of the term "environmental refugee" and instead refer to "populations at risk" because protection is both feasible and likely in many cases. Almost 100 million people are at risk in four countries alone: China, Bangladesh, Egypt, and Nigeria. In fact, Bangladesh is regarded as the most vulnerable country in the world because of coastal impacts (sea level rise and cyclones) as well as its reliance on outside assistance (e.g., 85% of Bangladesh's GDP is foreign aid). These results are based on present populations, and significant growth of populations at risk can be expected in the coming decades as sea level continues to rise and possibly accelerates.

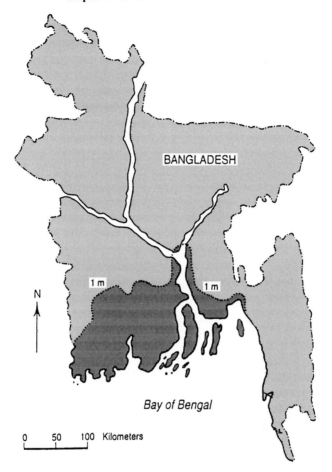

Figure 8.12   Land at risk in the Bangladeshi delta in response to a 1-m sea level rise (after Huq *et al.*, 1995).

Integrated delta management that includes upstream catchment areas should be instituted to minimize the vulnerability of coastal deltaic areas to sea level rise. Rivers that carry large sediment loads could at least theoretically be harnessed to raise land levels and hence counter the effect of sea level rise. This procedure has already been proven effective in the Yellow River delta, where the land surface has been raised several meters through controlled flooding (Han *et al.*, 1995). Similar strategies may be appropriate for other deltas worldwide. For example, controlled diversions of the Mississippi River in coastal Louisiana are currently being undertaken to counter rapid land loss (about 10 hectares per day) by providing both sediment and freshwater to preserve deltaic marshlands (Boesch *et al.*, 1994). The effectiveness of this vast experiment in terms of marshlands saved and restored versus cost is still not defined and remains controversial in many quarters. For example, the World Bank has a development plan to control river flooding in Bangladesh that would repeat the mistakes made by the U.S. Army Corps of Engineers over 50 years ago in coastal Louisiana. When environmental problems have long lead times in terms of damage realized versus the timing of the engineering project, it is often difficult to convince coastal residents or funding agencies of the eventual harmful consequences of actions which today can be beneficial (e.g., prevention of flooding in populated areas).

For some delta systems, sediment management is not possible because upstream dams have already removed the major source of material. Sediment bypassing of the Aswan High Dam on the Nile River in Egypt is probably not technically feasible because of insufficient river flow to carry the sediment from the dam to the delta (Milliman *et al.,* 1989). Therefore, only conventional engineering solutions, such as polderization as extensively practiced in the Netherlands, may make economic sense. Whatever strategies are adopted, long lead times for such alterations are required. In fact, sea level rise impacts, which are slowly realized but cumulative in effect, necessitate careful planning with a multidecadal to century time frame, which is difficult to institute in both industrialized and less developed countries.

Beyond the obvious ecological problems inherent in human manipulation of deltaic areas and the consequences of rising sea level, the prospect of massive numbers of persons drowning during high storm surges underscores the extremely high vulnerability of development on these low-lying lands. Many of the world's most populated deltas are located in the tropical storm belt, including Asia, India/Bangladesh, and coastal Louisiana in the United States. In the early 1970s, a single cyclone from the Bay of Bengal killed hundreds of thousands of people in the Ganges-Brahmaputra delta; recall that former Beatle George Harrison organized a special concert to aid the survivors. Lesser disasters in terms of sheer numbers occur every few years there, and the population push causes the survivors to take over the land from the dead. Even reliable casualty estimates are impossible to make since people swarm into the exposed land immediately after the storm surge crests, and they do not want to admit that they do not own the land they are now occupying. Relative sea level rise and the cutting down of mangroves, which formerly slowed and lowered the overall storm surge, have placed coastal Bangladesh even more at risk during the past century.

Severe flooding problems are not limited to developing countries. New Orleans, which is mostly below sea level, is protected by a circle dike designed to provide protection from a 100-year storm event. Andrew, a Category 4 hurricane in 1992, mercifully missed this legendary city, looping around it on three sides and not testing the sea defenses. Japan has several large urban areas, including parts of Tokyo, that are called "below zero level cities" because they now lie below mean sea level due to a long-term trend of high relative sea level rise. Typhoons and tsunamis put large populations located there in a perilous situation during these episodic events.

8.6   SMALL ISLAND VULNERABILITY

Coral reef atolls and reef islands, especially susceptible to even small rises of sea level, constitute the most vulnerable nations in the world. Some island states could cease to be habitable in the future (Roy and Connell, 1991). Most of these countries have limited resource bases and hence are ill equipped to handle existing environmental problems, such as explosive population growth, overdevelopment and pollution (Leatherman, 1997). These problems will only worsen as rising sea levels cause land submergence, beach erosion, increased storm flooding, higher water tables, salinity intrusion and overall reduced fresh water supply. Such changes will make these small land masses, many at or near existing sea level, less habitable for humans, resulting in off-island migration.

Small island states, typically less than 10,000 km^2 in area with approximately one-half million or fewer residents, vary by geography, social composition, political influences, economic priorities, physical make-up, and climatic conditions (Hess, 1990). Some small island states are essentially single islands (e.g., Sri Lanka and Barbados), while others are groups or archipelagoes of several (e.g., Tuvalu), hundreds (e.g., Tonga), or thousands (e.g., the Maldives) of islands. While some islands are mountainous (e.g., Seychelles) and contain active volcanoes (e.g., Montserratt), the most vulnerable islands are those consisting entirely of atolls and reef islands (e.g., Kirabati). These low-lying land masses are subject to submergence, erosion, and storm surge flooding. In addition, freshwater is often in scarce supply for human consumption and agricultural production because of limited groundwater.

Many small island countries would lose a significant part of their land area with a sea level rise of 1 m. For instance, the 1190 small islands that constitute the Republic of the Maldives have average elevations of 1 to 1.5 m above existing sea level (Pernetta, 1992). Submergence and erosion can convert many smaller islands to sandbars and significantly reduce the usable dry land on the larger, more populated islands. Saltwater intrusion and loss of the freshwater lens with higher sea levels may have equally severe limitations for human habitation (Roy and Connell, 1991). Rising sea level is already causing problems on the Marshall Islands, pushing salt water into groundwater supplies, making them useless for crops and threatening island drinking water. In some small atolls, the lens of freshwater lying above the salt water is now only a few centimeters thick (Leatherman, 1994).

Studies of Pacific islands have shown that there are many possible responses to sea level rise, involving the balance between reef growth and island accumulation or destruction (McLean and Woodroffe, 1993). Factors of importance include (IPCC, 1996) (1) island location within or beyond tropical storm belts, (2) composition being primarily of sand compared to coral rubble, (3) being anchored or not to emergent rock platforms, (4) presence or absence of natural physical shore protection in the form of beachrock, and (5) biotic protection—or its absence—in the form of mangrove or other strand vegetation. McLean and Woodroffe (1993) envisioned at least three different responses to sea level rise: (1) the Bruun response, (2) the equilibrium response, and (3) continued growth, which would result in (1) shore erosion and island shrinking, (2) redistribution of sediment on an island-wide basis, and (3) beach accretion and vertical build-up, respectively. More recently, Leatherman (1997) surveyed small island states in the Pacific Ocean, the Indian Ocean, and the Caribbean Sea and found a preponderance of erosional and other problems associated with sea level rise.

The characteristics of small island states present many constraints for sustainable development (Maul, 1996). Even without climate change and associated sea level rise, these nations will continue to experience increasing vulnerability to natural hazards because of anthropogenic influences including high population growth rates, overdevelopment, and continued exploitation of coastal resources (e.g., overfishing, sand mining of beaches) as well as the resulting pollution problems and decline of the resources that sustain their economies (e.g., tourist beaches and living coral reefs). The following list of characteristics of small island states defines their vulnerability:

- small size and limited arable land
- limited range of natural resources
- susceptibility to natural hazards (e.g., hurricanes and tsunamis)
- little biological diversity
- relative isolation and great distance to other markets
- extensive land/sea interface per unit area, making protective measures extremely expensive
- economies very susceptible to external shocks
- low resilience of a subsistence economy

The entirety of small islands is in their coastal zone, and sea level rise can affect virtually all the habitable area. Rising sea level will increase land loss through erosion of beaches and inundation of low-lying lands. Because of narrower and fewer sandy beaches, the tourist industry that generates a large part of the GDP in small island economies (15–18% in many Caribbean countries and 18% in the Maldives) will decline (UNESCO, 1994). Saltwater intrusion is especially a problem on coral reef atolls, where freshwater supplies are already limited. Because the population and economic activities are found primarily in the coastal zone, sea level rise threatens the governmental, industrial, tourist, energy, transportation, and communication infrastructure. In addition, as land is lost because of sea level rise, there will be an increase of outmigration from these islands.

Kosrae Island in the Federated States of Micronesia is prototypical of population centers and infrastructure along the coastal fringe. Here, schools are located at the water's edge, protected by a low coral rubble seawall (Fig. 8.13). The only hard-surface road on this island (the circle island highway)

Figure 8.13   A low seawall provides protection from saltwater flooding on Kosrae Island, Federated States of Micronesia.

was built just landward of the beach, and this newly constructed road is already experiencing erosion and failure (Fig. 8.14).

There are essentially two approaches that can be used to predict future erosion rates for sandy beaches: trend analysis based on historical shoreline positions, and Bruun's rule that long-term beach erosion is a very large multiple of the rate of sea level rise. Trend analysis involves the use of historical shoreline change data. Along the U.S. Atlantic and Gulf coasts, accurate map and aerial photographic data extend over a century, and computer-based mapping/geographic information system programs are used to obtain very detailed and accurate information on past shoreline behavior (Leatherman, 1983; Crowell and Leatherman, 1999). Island scientists, faced with a lack of sophisticated equipment and paucity of data for analysis, often rely upon point

Figure 8.14 The newly built and only hard-surface road on Kosrae, Micronesia is already subject to erosional problems.

measurements from photographs. Other evidence for long-term erosion is more bizarre. At Kosrae Island, Japanese military pillboxes and bunkers are now on the active beach face and subject to wave attack at high tide (Fig. 8.15); they had originally been built during World War II on upland areas, probably behind protective sand dunes. Abandoned concrete foundations of more recently constructed houses indicate progressive erosion at Malem, Kosrae Island (Fig. 8.16).

For point measurements from two sets of air photos, the longer the period of time recorded, the more reliable is any calculation of the average erosion rate (Crowell, *et al.,* 1993; Galgano, *et al.,* 1998). Extrapolations can be made into the future for the amount of erosion to be expected in response to accelerated sea level rise, assuming that the past rise in sea level accounted for the historic rate of loss. This assumes that the historical rate of sea level rise is known and that sand mining of the beach has not greatly contributed to the erosion rate. As already discussed, the Bruun rule is being used worldwide to project future shoreline position in response to sea level rise (Leatherman, 1991). Critical parameters are the depth of closure and width of the active profile (distance from the beach ridge to the depth of closure). These variables are difficult to estimate without good wave data (Nicholls *et al.,* 1995). In the case of small island states, wave information is generally not available. Furthermore, the Bruun rule is based on the concept of a totally sandy beach and nearshore profile, clearly not the case for a coral reef environ-

Figure 8.15   Japanese pillboxes now located in the intertidal zone are subject to wave attack during high tide (Malem shoreline, Kosrae, FSM).

Figure 8.16 Concrete foundations of "western" style houses lie abandoned in the intertidal zone, but the new house is located only a few meters landward of the one previously abandoned.

ment. Therefore, the Bruun model loses physical meaning to a considerable degree for these situations.

Two methods of preventing land loss are beach nourishment and construction of seawalls. Each approach has its limitations in terms of application on small island states. Beach nourishment is the preferred alternative for dealing with the beach erosion problem along the U.S. coasts, but is rarely used on island states (Leatherman, 1996). The major problem is the availability of sand-sized material. Most islands have no major river systems to act as a source of sand. In addition, there are no major navigation channels that can be dredged as a source of sand for beach fill projects. This lack of sand sources usually makes beach nourishment an uneconomical option. Indeed, sand is often mined from the beach face as the primary source for construction material for concrete-block houses. In Grenada, for example, sand mining for construction purposes is one of the main factors contributing to beach degradation (Isaac, 1997).

Shoreline stabilization has been used widely to counter ongoing beach erosion on island beaches (Fig. 8.17). Many of these projects have been ill-conceived, causing as much or more damage than benefit. Hard shoreline engineering has been oversold as a panacea to coastal erosion problems. The consequences are all too clear in places like Sri Lanka, where beautiful white sandy beaches have been squeezed out of existence. Seawalls also encourage unwise development practices by allowing hotels and infrastructure to be constructed too close to an eroding shore. For many small island states, tourism is the major source of hard currency. Beaches are the prime recreational draw and support much of the infrastructure. Therefore, building setbacks must be

Figure 8.17   Hard stabilization was necessary to protect the airport at Majuro, Marshall Islands.

instituted to provide the buffer zone needed under existing conditions as well as to provide a measure of protection as the seas rise and beach erosion rates accelerate in the coming decades.

8.7   THE CHESAPEAKE BAY: A CASE STUDY

The Chesapeake Bay (Fig. 8.18) serves as an ideal laboratory for investigating the physical effects of accelerated sea level rise and exploring the range of impact and response strategies. The region is characterized by diverse coastal environments, from the high cliffs of the Western Shore to the low-lying coastal plain on the Eastern Shore. The Bay's shoreline, many 1000s of kilometers in length, is subject to the entire panoply of sea level rise impacts: erosion, inundation of low-lying lands, wetland loss, saltwater intrusion into aquifers and surface waters, and increased flooding and storm damage. Bay islands that supported entire villages are disappearing, and vast expanses of salt marsh are drowning. With such a diversity of landforms and ecosystems responding to sea level rise, the Chesapeake Bay serves as a microcosm of threatened coastal areas worldwide.

Coastal land loss in the Chesapeake Bay region is a major economic and political issue today because more and more people are moving close to the water's edge. At the same time, sea level in the area has been rising at a rapid

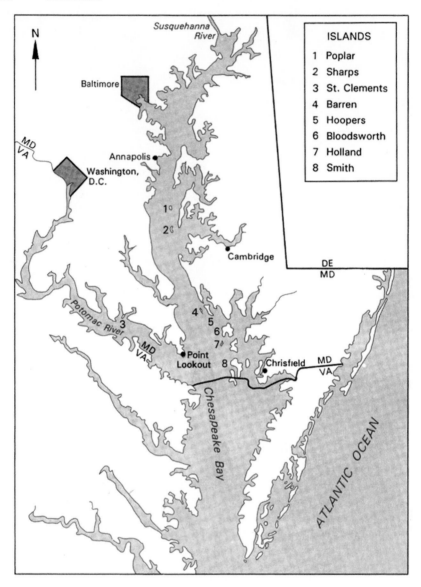

Figure 8.18   Islands of the Chesapeake Bay in Maryland.

rate—about 0.35 m during the last century. We are provided here with a prototypical example of the collision course between social and environmental trends.

The Chesapeake Bay coastline responds to sea level rise in two different ways: submergence and erosion. While submergence and erosion are both evident on the Bay's Eastern Shore, cliff retreat through erosion is dominant on the Western Shore. Land loss is not a recent phenomenon. It has occurred gradually at least since European settlement in response to a more slowly rising sea level at that time, with map evidence indicating substantially increased rates of loss in the latter 19th and 20th centuries (Kearney and Stevenson, 1991; Leatherman *et al.,* 1995b) in response to higher rates of sea level rise. Because of the early exploration and settlement of the Chesapeake Bay region, excellent historical data are available. For instance, Kent Island was settled in 1631, and original Royal land charters and tax records of the Chesapeake Bay islands have been kept since that time. Maps, charts, and photos obtained from historical societies, museums, and the National Archives document the changes over the centuries. Land loss information can also be obtained from newspaper accounts of storm events or the demise of an island. To illustrate the varying impacts of sea level rise, morphologically differing areas are considered below in some detail.

8.7.1   Western Shore of Maryland

The Western Shore, the cliffed shoreline of Maryland, is composed of sediments ranging from loose sand to blue marine clays that tower up to 30 m above the Bay water level (Fig. 8.19). These impressive cliffs, almost vertical in places, are maintained by wave undercutting at their toe. Retreat rate range from 0.1 m to almost 2 m per year based on comparisons of historical maps and charts (Leatherman *et al.,* 1995b).

Waves, particularly during storms, are literally eating away part of Maryland's history, as bits and chunks of sediment fall along the retreating cliff face. St. Clement's Island on the Potomac River, where the Colonists first landed in 1634, is rapidly disappearing. When first settled, it covered over 160 hectares and was described as "thickly wooded with cedars, sassafras and nut trees, with herbs and flowers everywhere"; today it has shrunk through cliff erosion to about 16 hectares with few remaining trees. In spite of this rapid land loss, the U.S. government built a lighthouse on the southern tip of the island in 1853. The lighthouse remained active until the 1920s; eventually erosion caused its collapse into the Potomac River.

Jutting out into the sea where the Potomac River empties into the Chesapeake Bay lies Point Lookout, one of the most scenic spots on the Bay. The original land grant of 1186 hectares originated from Lord Baltimore in 1634. During the Civil War, over 10,000 Confederate soldiers were confined here, but nearly all the remnants of this Civil War–age encampment have succumbed to the sea. As much as 50% of Point Lookout has eroded away in the past 100 years, based on map comparisons (Fig 8.20). While erosion has not been rapid by Baywide standards, a loss of a meter per year along the eroding edge can eventually claim considerable real estate.

Figure 8.19 Erosion of the Western Shore cliffs along the Chesapeake Bay has resulted in the loss of another house in the Randle Cliffs area.

Coastal land loss can be prevented, but the price can be high depending on wave energy and currents. The historic towns and present-day ports of Annapolis and Baltimore have been effectively fortified against sea level rise impacts. These harbors are naturally protected from the larger waves along the main stem of the Bay by their location. Such urban areas receive protection through the construction and maintenance of bulkheads and revetments.

8.7.2 Eastern Shore of Maryland

The flat and low-lying Eastern Shore contrasts sharply with the high eroding bluffs of the Western Shore. While there is erosion along the Bay edge, the

Figure 8.20   Map comparisons of Point Lookout, showing the present-day shoreline superimposed on a Civil War–era map (courtesy of National Archives).

primary response to rising sea levels on the Eastern Shore is submergence of upland areas and drowning of coastal marshes.

The Eastern Shore is the site of one of the largest and most important expanses of coastal wetlands along the U.S. mid-Atlantic coast. The Blackwater National Wildlife Refuge near Cambridge, Maryland, is heavily used by Canada and snow geese and numerous species of ducks. Unfortunately, as aerial photographs reveal (Fig. 8.21), over one-third of the total marsh area, about 2,000 hectares, was lost between 1938 and 1988, and the trend is continuing. Marsh losses have also been documented at other wildlife management areas in the Chesapeake and Delaware Bays. During the past 50 years an imbalance has developed so that relative sea level rise is occurring significantly faster than deposition on the marsh surface, eventually resulting in saltwater intrusion, waterlogging and plant demise. Intertidal salt marshes can only adapt to relatively moderate rates of sea level rise; rapid increases in sea level (~5 mm/yr, only a little more than the present rate in the middle Atlantic region) can literally drown most of the Bay's wetlands, converting them to shallow water areas. As sea levels continue to outpace the ability of the marsh to maintain elevation, the magnitude of the total loss will increase (Leatherman *et al.,* 1995b).

While existing salt marshes are sustaining heavy losses, which will only accelerate through time, the real hope for the future is the new marshes developing at the shore edge. However, landward penetration of brackish water plants is viewed by Eastern Shore residents as a loss of good agricultural

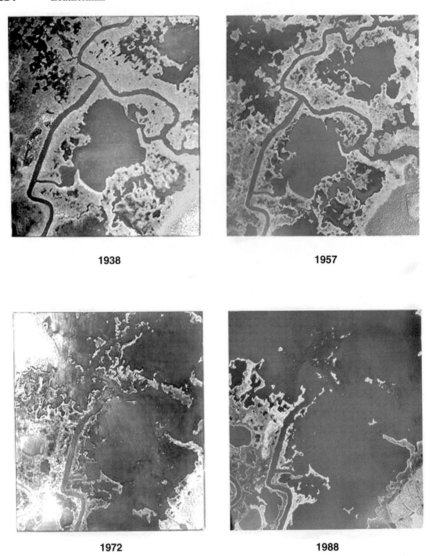

1938

1957

1972

1988

Figure 8.21 Progressive drowning and loss of coastal marshes in Blackwater National Wildlife Refuge near Cambridge, Maryland, in response to a high relative sea level rise.

land. While shrinking of arable lands will undoubtedly reduce the farmer's yield of corn and other crops, the loss is not significant in the larger sense of the region. Ample space is available elsewhere for agricultural activities, and defensive measures such as bulkheads are not justified because of the high cost of protecting this low-lying land.

The real fate of new marshes is determined by increased development. As the Bay's shore continues to urbanize, the perceived benefits of development may justify the costs. Bulkheads will then be built to line the high ground, and marshes will literally be squeezed out of existence with sea level rise (Fig. 8.22).

8.7.3 Bay Island Communities

The human dimension of sea level rise impacts is best exemplified by an examination of land loss of the Chesapeake Bay islands. Sizeable islands have been substantially reduced in size or even drowned as shown on successive 19th- and 20th-century maps and charts (Kearney and Stevenson, 1991; Leatherman, 1992). Sharp decreases in island area led to widespread abandonment of settlements on many of the islands in the first decades of this century (Table 8.5). Depopulation of islands has occurred as rising sea levels have caused progressive erosion, submergence, or both, eventually eliminating human habitation, especially in the wake of a hurricane.

The Bay islands, composed of fine-grained clay deposits, are artifacts of the changing course of the Susquehanna River. Their erosion and submergence are natural geomorphic phenomena driven by rising sea level. The highest points on the Bay islands are ridges, but even here elevations of 3 m are not attained. Most of the land area exists from near sea level to less than 1 m. The smallest islands of less than 400 hectares, which also tended to be the lowest, have fared worse as sea level has risen and much dry land has been

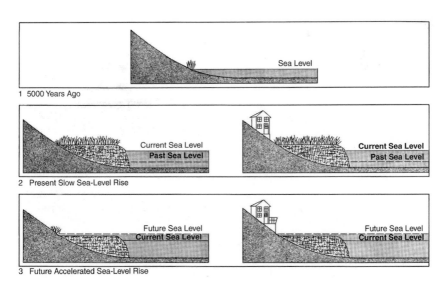

Figure 8.22 Evolution of coastal wetlands in response to sea level rise. Note that future projections of existing houses with bulkheads can result in a total loss of wetlands.

Table 8.5

Historic Size Comparison of Chesapeake Bay Islands

Island	Historic acreage	(date)	Recent acreage	(date)	% Lost	Comments
Poplar	1400	(1670)	125	(1990)	91	Abandoned in 1930
Sharps	890	(1660)	0		100	Drowned in 1962
St. Clements	400	(1634)	40	(1990)	90	Abandoned in 1920s
Barren	700	(1664)	250	(1990)	64	Abandoned in 1916
Hoopers	3928	(1848)	3085	(1942)	21	Submerging
Bloodsworth	5683	(1849)	4700*	(1973)	17	Submerging
Holland	217	(1668)	140*	(1990)	35	Abandoned in 1922
Smith	11,033	(1849)	7825*	(1987)	29	Submerging

*Mostly marshy land.

lost. While the much larger islands of Smith and Tangier still support towns, the last permanent residents have left Holland and Poplar Islands, and Sharps Island has vanished.

The original size of Poplar Island is estimated at 800 hectares based on the outline of the existing sand shoal now surrounding the island. The first recorded sale transaction in the late 1670s reported 560 hectares. During the War of 1812, the British fleet occupied Poplar, as they had during the American Revolution. In 1848 the first truly accurate map of this island was produced by the U.S. Survey of the Coast (Fig. 8.23). The island was greatly reduced in size from the 1600s, but it still measured roughly 316 hectares. Erosion and

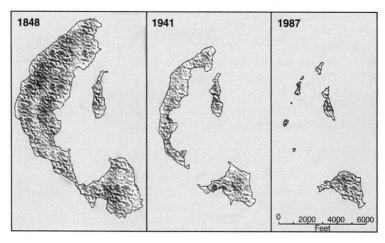

Figure 8.23 Historical shoreline changes of Poplar Island, Maryland, which was the favorite retreat of President Franklin D. Roosevelt.

submergence continued to take their toll and this single island ultimately split into a main island and several smaller islands.

Around 1900 at the height of settlement, about 15 families totaling 70 to 100 people lived on the Poplar Islands. The main island was still large enough to support the small town of Valliant. This town included a general store, post office, school, church, and a sawmill. The community flourished until the 1920s when serious erosion became evident, and most islanders abandoned their homes as the island began to disappear. Local activity turned to hunting and moonshining. Both Presidents F. D. Roosevelt and H. S. Truman were visitors to the island in its days as a hunting club. The last frequent visitors to the vanishing island were a family who commuted from Philadelphia on weekends with their children and guests, and subsequently donated the island to the Smithsonian Institution for use as a research site. Relative sea level rise and the resulting erosion caused Poplar's demise. The island is slipping away at the rapid rate of more than 4 meters a year and will probably disappear in a few years at present rates of land loss unless drastic action is taken.

Recently, barges have been sunk, and sand dredged from Baltimore Harbor has been imported to resurrect Poplar Island as a bird sanctuary. The total cost to resurrect the island is currently estimated at an incredible $427 million. The effort to restore disappearing Poplar Island in the Chesapeake Bay represents a new and positive direction in dredging operations—the beneficial use of clean dredge spoil to enhance the environment. Underway since 1996, the project is one of several selected by the Maryland Port Administration and the U.S. Army Corps of Engineers (Baltimore District) to maintain and improve the 126 miles of Federal navigation channels serving the Port of Baltimore and keeping it competitive as ships grow bigger and draw more water.

Known locally as the "sinking island," Bloodsworth Island's arable land has given way to brackish water over the centuries, with corn fields reverting to salt marshes. The recorded history of Bloodsworth Island indicates the changing uses of the land over time in response to slowly rising sea level. The earliest map (1637) shows the area as a series of small islets surrounded by a fringe marsh. At least 20 hectares were under cultivation with a few small houses and fruit trees also present (Leatherman, 1995). The first accurate map of the area, produced in 1849 by the U.S. Survey of the Coast, shows seven structures on small upland tracts. By this time the island was dominated by wetlands, but brick rubble and building foundations have been found buried about 15 cm down. In response to the rising water table and higher salinity of the soil, the land became less suitable for permanent occupancy. By the 1920s the island was abandoned as tidal flooding caused the land to be too wet and salty for cultivation. Because of its somewhat protected position, shallow offshore water, and marshy, erosion-resistant substrate, the total land area has changed little in size, but it is now covered with salt marsh vegetation. The large expanse of marshes at Bloodsworth can be self-maintaining to some extent under slowly rising sea level, but an accelerated rate of rise will drown these coastal wetlands with no chance for migration. Therefore, this island has a limited future in response to predicted sea level rise.

8.8   CONCLUSIONS

Sea level rise is the most certain consequence of global warming, and the impacts are already evident along the world's coast. In economic terms, beach erosion is the most serious problem as tourist beaches diminish in size or are lost over time. It is estimated that 80–90% of U.S. sandy beaches are currently experiencing erosion, and the cost of beach restoration through sand pumping is quite high. Therefore, beach nourishment makes economic sense only for the most developed areas ("cities on the beach" like Miami Beach, Florida, and Ocean City, Maryland).

The most vulnerable areas to sea level rise impacts in human terms are deltas and small island nations. Millions of people live in low-lying, deltaic areas in some of the poorest countries in the world (e.g., Bangladesh, Egypt, Nigeria, and China). The Chinese fortunately have the will and means to protect much if not all of their coastal population at risk. For instance, higher and more extensive dikes have recently been completed to protect the populace of Shanghai. By contrast, the residents of Kirabati, a country built on low-lying coral reef atolls, cannot afford to protect even the capital city, much less the many outlaying islands. Small islands are problematic to defend against beach erosion and flooding because of their long shoreline versus small areal extent.

Possession of the economic resources to accommodate sea level change is the critical question. Table 8.1 lists the adaptation/protection costs for several less-developed countries. Bangladesh is a classic case where major impacts from accelerated sea level rise will affect a large number of people, but where economic development cannot be expected to produce the resources to support any large-scale strategies for mitigation without outside help. Compounding this problem are proposed polder construction project which, while contributing to increased rice production in the short term, may exacerbate land loss in the delta versus sea level rise over the long term (Stevenson and Kearney, 2000). However, it would be a mistake to believe that only the developing nations face a lack of resources in adjusting to sea level rise. Again, the Chesapeake Bay can serve as an example. With 6000 to 8000 miles of shoreline (depending on how the shoreline is determined) and burgeoning development along its breadth, the costs of extensive shore protection for the Bay are prohibitive. Recent estimates by the U.S. Army Corps of Engineers place the cost of typical shore erosion protection measures at between $400 to almost $1000 per linear foot (0.3 m), depending on the level of wave activity (U.S. Army Corps of Engineers, 1990). A quick calculation shows that the cost of extensive shore protection in the Chesapeake Bay could easily exceed the economic damages from Hurricane Andrew, the most costly (approx $25 billion) Atlantic hurricane of the 20th century. Moreover, many of the measures will prove merely temporary as sea level continues to rise, needing reconstruction or replacement in a few decades.

Thus, it is clear that cost places an extraordinary burden on coastal modification as a policy for accommodating future sea level rise even for the wealthiest of nations. The alternative is effective coastal planning, with strategies for limiting the impact of sea level rise, whether they are land use policies, tax code incentives (or disincentives), or a mix of these and other approaches. Nevertheless, as the history of the environmental movement in the West has shown, implementation of any comprehensive policy for remediation of the sea level risk will entail considerable effort at education of the public to the problem and the costs of being unprepared. The issue of global warming has received continual coverage in most electronic and print media of western nations for over a decade, but whether the public is fully cognizant of its implications is open to question. With respect to the threats accelerated sea level rise poses for the world's coasts, the answer is mixed. Every hurricane season the U.S. media highlight the risks of coastal erosion and flooding and, increasingly, link global warming and sea level rise. However, coastal development continues. For the United States, this trend has resulted in ever greater costs for storm damage, especially from hurricanes. Even without an increase in the frequency of tropical storms, as sea levels rise and shorelines retreat, the risk of flooding becomes ever greater. It is hard to conceive of the devastation a storm with the strength of Hurricane Camille (Category 5, 1969, with sustained winds of 170 mph) would wreak today along the U.S. Gulf Coast. The burgeoning coastal development during the intervening three decades since this storm, coupled with half a meter of submergence around the Mississippi Delta, has created the potential for a major disaster.

There is hope that education on the risks implicit to coastal communities in a scenario of rapidly rising sea levels could make a difference. In order to be effective, the *real* potential damages must be conveyed. Recent attempts by the academic and professional coastal scientists worldwide to address the "hidden" costs implied by coastal hazards are encouraging (Heinz Center, 2000).

For coastal wetlands, the future is not certain. Though the value of coastal wetlands is now recognized, ranging from food and nurseries for fish and shellfish, to provision of over-wintering habitat, losses of coastal wetlands continue in the United States despite implementation of a policy of "no net wetland loss" over a decade ago. One of the areas of the most dramatic losses of coastal wetlands is in southern Louisiana. About 10 hectares of marshlands are lost there *per day,* due to the high relative rate of sea level rise of about 1 cm/yr (largely due to land subsidence, in part caused by the withdrawal of subsurface water, oil, and gas). High rates of coastal marsh loss have also been reported in the well-studied Maryland Chesapeake Bay where there is no petroleum production, but geologic causes (see Chapter 4) and possibly overpumping of groundwater are responsible for substantial land losses. Here, like other regions of widespread marsh loss, whatever the impacts of human activities, the primary underlying factor is a relatively high rate of sea level rise exceeding the rate of sedimentary accumulation on the wetland surface.

Confronting the issue of coastal wetland loss will require better information on wetland status than is currently available. Such data will have to address the susceptibility of the marsh to future loss from sea level rise with respect to geomorphic setting, marsh type, and whatever degradation has already occurred. With respect to mitigation of damage from sea level rise, the solutions are complex. There has yet to emerge a reasonable, cost-effective strategy for reversing the course of the marsh loss cycle, particularly in its later stages. For large deltaic systems, there may be tractable solutions. Marsh re-creation may be the only answer, but with the caveat that what is being created probably will not replace the variety of complex denitrification and sulfate reduction mechanisms inherent in mature marsh systems that evolved over millennia.

Ultimately, there is no silver lining to the sea level rise story. The hope is that people will plan for its consequences because the rate of annual rise is so small that impacts take decades to be realized. But this slow rise is almost insidious—it is often ignored until the problem is brought home by a major storm event. Unfortunately, society is much more prone to respond to a crisis situation. Environmental problems with long lead times (such as rising sea level) are often given inadequate attention by citizens and politicians alike.

REFERENCES

Bird, E. C. F. (1976). Shoreline changes during the last century. *Proceedings of the 23rd International Geographical Congress,* Pergamon, Elmsford, NY.

Bird, E. C. F. (1985). *Coastline Changes—A Global Review.* Wiley, Interscience, Chichester.

Boesch, D. F., Josselyn, M. N., Mehta, A. J., Morris, J. T., Nuttle, W. K., Simenstad, C. A., and Swift, D. J. P. (1994). Scientific assessment of coastal wetland loss, restoration and management in Louisiana. *J. Coastal Res.* Special Issue **20.**

Bruun, P. (1962). Sea level rise as a cause of shore erosion. *J. Waterways Harbors Div.* **88,** 117–130.

Bruun, P. (1983). Review of conditions for uses of the Bruun Rule of erosion. *Coastal Eng.* **7,** 77–89.

Carter, W. E., Shrestha, R. L., and Leatherman, S. P. (1998). Airborne laser swath mapping: Applications to shoreline mapping. *Proc. INS MAP 98 Conf.* Melbourne, Australia.

Crowell, M., and Leatherman, S. P. (eds.) (1999). Coastal Erosion Mapping and Management, *J. Coastal Res. Special Issue* **28.**

Crowell, M., Leatherman, S. P., and Buckley, M. K. (1991). Historical shoreline change: Error analysis and mapping accuracy. *J. Coastal Res.* **7,** 839–852.

Crowell, M., Leatherman, S. P., and Buckley, M. K. (1993). Shoreline change rate analysis: Long term versus short term data. *Shore and Beach* **61,** 13–20.

Douglas, B. C. (1991). Global sea level rise. *J. Geophys. Res.* **96,** 6981–6982.

Douglas, B. C. (1997). Global sea rise: A redetermination. *Surveys Geophys.* **18,** 279–292.

Ellison, J. C., and Stoddart, D. R. (1991). Mangrove ecosystem collapse with predicted sea-level rise: Holocene analogues and implications. *J. Coastal Res.* **7**, 151–165.

Galgano, F. (1998). *Geomorphic Analysis of Modes of Shoreline Behavior and the Influence of Tidal Inlets on Coastal Configuration.* Ph.D. dissertation, University of Maryland, College Park.

Galgano, F., Douglas, B. C., and Leatherman, S. P. (1998). Trends and variability of shoreline position. *J. Coastal Res.* Special Issue **26**, 282–291.

Gehrels, W. R., and Leatherman, S. P. (1989). Sea level rise—animator and terminator of coastal marshes: An annotated bibliography on U.S. coastal marshes and sea level rise. Vance Bibliographies Public Administration Series P 2634.

Hallermeier, R. J. (1981). A profile zonation for seasonal sand beaches from wave climate. *Coastal Eng.* **4**, 253–77.

Han, M., Hou, J., Wu, L., Liu, C., Zhao, G., and Zhang, K. (1995). Sea level rise and the North China coastal plain: A preliminary analysis. *J. Coastal Res.* Special Issue **14**, 132–150.

Hands, E. B. (1983). The Great Lakes as a test model for profile response to sea level changes. In *CRC Handbook of Coastal Processes and Erosion,* pp. 167–189. CRC Press, Boca Raton, FL, 167–89.

Heinz Center (2000). *The Hidden Costs of Coastal Hazards.* Island Press, Washington, DC.

Hess, A. (1990). Overview: Sustainable development and environmental management of small islands. In *Sustainable Development and Environmental Management of Small Islands* (W. Beller, P. D'Ayala, and P. Hein eds.), UNESCO, Paris, 3–14.

Huq, S., Ali, S., and Rahman, A. (1995). Sea level rise and Bangladesh: A preliminary analysis, *J. Coastal Res.,* Special Issue **14**, 44–53.

IPCC (1992). *Global Climate Change and the Rising Challenge of the Sea.* Coastal Zone Management Subgroup, Intergovernmental Panel on Climate Change Working Group III, Rijkswaterstaat, The Netherlands.

IPCC (1996). *Coastal Zones and Small Islands.* Coastal Zone Management Subgroup, Intergovernmental Panel on Climate Change Working Group II, Rijkswaterstaat, The Netherlands.

Isaac, C. (1997). Sand mining in Grenada: Issues, challenges and decisions relating to coastal management. In G. Cambers (ed.), *Managing Beach Resources in the Smaller Caribbean Islands* UNESCO, Mayaguez, Puerto Rico, 69–76.

Kearney, M. S., and Stevenson, J. C. (1991). Island land loss and marsh vertical accretion rate: Evidence for historical sea level changes in Chesapeake Bay. *J. Coastal Res.* **7**, 403–415.

Leatherman, S. P. (1983). Shoreline mapping: A comparison of techniques. *Shore and Beach* **51**, 28–33.

Leatherman, S. P. (1986), *Effects of Sea level Rise on Coastal Ecosystems.* U.S. Senate Publ. 99–723, Washington, DC. 141–153.

Leatherman, S. P. (1991). Modeling shore response to sea level rise on sedimentary coasts. *Prog. Phys. Geog.* **14**, 447–464.

Leatherman, S. P. (1992). Coastal land loss in the Chesapeake Bay region: An Historical analogy approach to global climate analysis and response. In *The Regions and Global Warming: Impacts and Response Strategies* (J. Schmandt, ed.), pp. 17–27. Oxford Univ. Press.

Leatherman, S. P. (1994). Rising sea level and small island states. *Ecodecision* **11**, 53–54.

Leatherman, S. P. (1996). Shoreline stabilization approaches in response to sea level rise: U.S. experiences and implications for Pacific islands and Asian nations. *J. Water Air Soil Pollution* **92**, 149–157.

Leatherman, S. P., (ed), (1997). *Island States At Risk: Global Climate Change, Population and Development. J. Coastal Res.* Special Issue **24**.

Leatherman, S. P., and Nicholls, R. J. (1995). Accelerated sea level rise and developing countries: An overview. *J. Coastal Res.* Special Issue **14**, 1–14.

Leatherman, S. P., Nicholls, R. J., and Dennis, K. C. (1995a). Aerial videotape-assisted vulnerability analysis: A cost-effective approach to assess sea level rise impacts. *J. Coastal Res.* Special Issue **14**, 15–25.

Leatherman, S. P., *et al.* (1995b). *Vanishing Lands: Sea Level, Society and the Chesapeake Bay.* U.S. Fish & Wildlife Service, Annapolis, MD.

Leatherman, S. P., Zhang, K., and Douglas, B. C. (2000). Sea level rise shown to drive coastal erosion. *EOS Trans. AGU* **81** (6), 55–57.

McLean, R. F., and Woodroffe, C. D. (1993). Vulnerability assessment of coral atolls: The case of Australia's Cocos (Keeling) Islands. *Proc. IPCC/WCC 93 Eastern Hemisphere Preparatory Workshop,* Tsukuba, Japan, 99–108.

Maul, G. A. (1996). Small islands: Marine science and sustainable development. Coastal and Estuarine Studies, America Geophysical Union, Washington, DC.

Milliman, J. D., Broadus, J. M., and Gable, F. (1989). Environmental and economic implications of rising sea level and subsiding deltas: The Nile and Bengal examples. *Ambio* **18,** 340–5.

Morton, R. A., Paine, J. G., and Gibeaut, J. C. (1994). Stages and durations of post storm beach recovery, southeastern Texas coast, U.S.A. *J. Coastal Res.* **10** (4), 884–908.

National Research Council (1990). *Managing Coastal Erosion.* National Academy Press, Washington, DC.

Nicholls, R. J. (1995). Synthesis of vulnerability studies. In *Preparing to meet the Challenges of the 21st Century. Proc. World Coast Conference,* The Hague, Netherlands.

Nicholls, R. J., and Leatherman, S. P. (eds.) (1995a). *Potential Impacts of Accelerated Sea Level Rise on Developing Countries. J. Coastal Res.* Special Issue **14.**

Nicholls, R. J., and Leatherman, S. P. (1995b). Sea level rise. In *As Climate Changes: Impacts and Implications,* pp. 92–123. Cambridge Univ. Press.

Nicholls, R. J., Leatherman, S. P., Dennis, K. C., and Volonte, C. R. (1995). Impacts and responses to sea level rise: Qualitative and quantitative assessments. *J. Coastal Res.* Special Issue **14,** 26–43.

Peltier, W. R., and Jiang, X. (1997), Mantle viscosity, glacial isostatic adjustment and the eustatic level of the sea. *Surveys Geophys.* **18,** 239–277.

Pernetta, J. C. (1992), Impacts of climate change and sea level rise on small island states: National and international responses. *Global Environmental Change* **2,** 19–31.

Roy, P., and Connell, J. (1991). Climatic change and the future of atoll states. *J. Coastal Res.* **7,** 1057–1075.

SCOR (1991). The response of beaches to sea level changes: A review of predictive models. *J. Coastal Res.* **7,** 895–921.

Snedaker, S. C., Meeder, J. F., Ross, R. S., and Ford, F. G. (1994). Discussion of Ellison and Stoddart. *J. Coastal Res.* **10,** 497–498.

Stevenson, J. C., Ward, L. G., and Kearney, M. S. (1986). Vertical accretion in marshes with varying rates of sea level rise. In *Estuarine Variability* (D. Wolf ed.), Academic Press, New York, 241–260.

Stevenson, J. C., and Kearney, M. S. (2000). Sea level, land subsidence and groundwater extraction impacts on coastal wetlands: How vulnerable is Bangladesh in the 21st century, and what can managers do to promote sustainability? *The International Seminar on Sea Level Rise and Sustainable Development in Bangladesh,* Dhaka, Bangladesh, in press.

Stive, M. J. F., De Vriend, H. J., Nicholls, R. J., and Capobianco, M. (1992). Shore nourishment and the active zone: A time scale dependent view. *Proc. 23rd Coastal Engineering Conference,* American Society of Civil Engineers, New York, 2464–2473.

Theodossiou, E. I., and Dowman, I. J. (1990). Heightening Accuracy of Spot. *Photogram. Eng. Rem. Sensing* **56,** 1643–1649.

Titus, J. G., Park, R. A., Leatherman, S. P., Weggel, J. R., Green, M. S., Mausel, P. W., Brown, S., Gaunt, C., Trehan, M., and Yohe, G. (1991). Greenhouse effect and sea-level rise: Potential loss of land and the cost of holding back the sea. *Coastal Management* **19,** 171–204.

UNESCO (1994) *Island Agenda: An Overview of UNESCO's Work on Island Environments, Territories, and Societies.* Paris.

Vellinga, P., and Leatherman, S. P., (1989). Sea level rise, consequences and policies. *Climatic Change* **15,** 175–189.

Woodroffe, C. D. (1990). The impact of sea level rise on mangrove shorelines. *Prog. Phys. Geography* **14,** 483–520.

Wright, L. D. (1978). River deltas. In *Coastal Sedimentary Environments* (R. A. Davis, ed.), pp. 5–68. Springer Verlag, New York.

Zhang, K. (1998). *Twentieth Century Storm Activity and Sea Level Rise Along the U.S. East Coast and Their Impact On Shore Line Position.* Ph.D. dissertation, University of Maryland, College Park.

Index

International Geophysics Series

EDITED BY

RENATA DMOWSKA
Division of Engineering and Applied Science
Harvard University
Cambridge, Massachusetts

JAMES R. HOLTON
Department of Atmospheric Sciences
University of Washington
Seattle, Washington

H. THOMAS ROSSBY
Graduate School of Oceanography
University of Rhode Island
Narragansett, Rhode Island

* Out of Print